AF556881

Reviews of Physiology, Biochemistry and Pharmacology 136

Springer
Berlin
Heidelberg
New York
Barcelona
Hong Kong
London
Milan
Paris
Singapore
Tokyo

Reviews of

136 Physiology Biochemistry and Pharmacology

With 12 Figures and 14 Tables

ISSN 0303-4240
ISBN 3-540-65151-9 Springer-Verlag Berlin Heidelberg New York

Library of Congress-Catalog-Card Number 74-3674

Production: PRO EDIT GmbH, D-69126 Heidelberg
SPIN: 10695328 27/3136-5 4 3 2 1 0 – Printed on acid-free paper

Contents

Indexed in Current Contents

Immunomodulation by Colony-Stimulating Factors

Thomas Hartung

University of Konstanz, Biochemical Pharmacology, D-78457 Konstanz

Contents

Introduction

The immune system has the task of warding off the continuous daily onslaught of microorganisms, thus protecting the body from infections. The first line of defense mobilizes macrophages, polymorphonuclear granulocytes, natural killer cells and cytotoxic lymphocytes. Upon activation, these directly attack and destroy foreign bodies. An inflammatory response with the characteristic symptoms of local hyperthermia, swelling, redness and pain is the result.

Cytokines and lipid mediators constitute the signals which enable communication between the different leukocyte populations and the coordination of defense of major organs. In this way, a response overshoot, which would result in tissue or organ damage, can be prevented. The intricate regulation of the immune system by the vast variety of mediators will probably remain largely a mystery owing to the high level of redundancy, i.e. many of the messengers differ only slightly in the responses they provoke; and to the pleiotropy of the mediators, i.e. the same molecules are employed as mediators in a variety of different situations throughout the body. The colony-stimulating factors (CSF), endogenous cytokines now cloned and commercially available, bring about definite, reproducible immunomodulatory effects in vitro and in vivo, including the proliferation of selected leukocyte populations and the intensification of their mature functions.

This review will examine the immunomodulatory properties of granulocyte colony-stimulating factor (G-CSF) and granulocyte-macrophage colony-stimulating factor (GM-CSF), which have been studied most extensively up until now, as well as monocyte colony-stimulating factor (M-CSF) and interleukin-3 (IL-3, multi-CSF). Although these four cytokines mediate highly specific, overlapping biologic activity, their amino acid sequence dissimilarities within one species indicate that they are not derivatives of a common evolutionary ancestral regulatory molecule (Metcalf 1986). Furthermore, the homology between the human and murine genes varies between the CSF, consequently crossreactivity is not homogenous (Table 1).

Table 1. Homology and crossreactivity of human and murine CSF

CSF	homology hu : mu	crossreactivity
G-CSF	73%	yes
GM-CSF	56%	no
M-CSF	82%	hu→mu
IL-3	29%	no

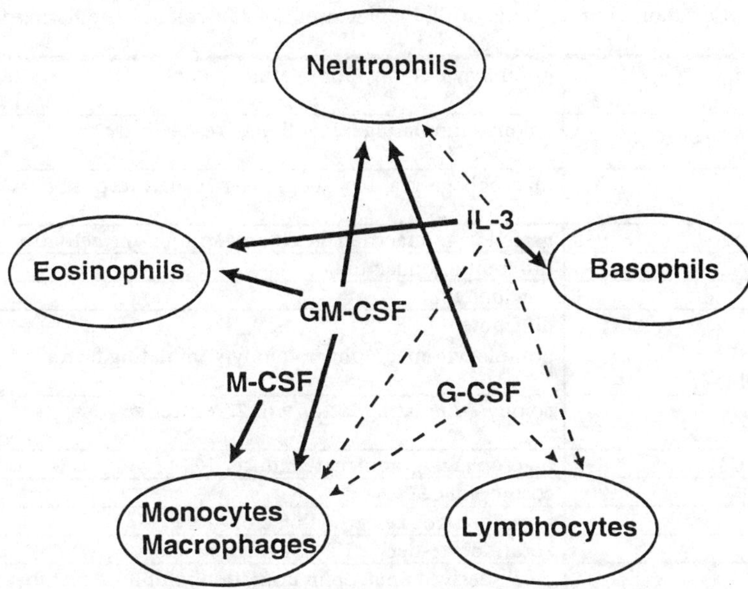

Fig. 1. The role of the four colony-stimulating factors in hematopoesis. The diagramm summarizes the hematopoetic effects of G-CSF, GM-CSF, M-CSF and IL-3. Bold arrows refer to strong and dashed arrows to weak stimulatory activities

General aspects of the molecule, its production, its receptor and its role in hematopoiesis (Fig. 1) will be discussed briefly for each of the CSF. Then, the immunomodulatory effects on the various leukocyte populations will be addressed. Since anti-infectious defense is accomplished by a joint effort of all immune cells, a separate chapter is devoted to the overall effect of the CSF on the course of infectious diseases. The final part of each section will discuss the emerging clinical potentials arising from the immunomodulatory activities of the CSF. If not stated otherwise, all data refer to the human system.

A number of biological activities, which provided names for a variety of putative mediators, are now attributed to the four pertinent CSF. Table 2 lists some examples of mediators found to be identical with these CSF. Table 3 provides a compilation of the bioassays established for the CSF discussed in this review.

Table 2. Common abbreviations of the names used for the colony-stimulating factors

G-CSF ; CSF-G	granulocyte colony-stimulating factor
CSF-β ; CSF-3	colony-stimulating factor β or 3, respectively
MGI-1G ; MGI-2	macrophage-granulocyte inducer 1G or 2 resp.
G/M-CSA	granulocyte-macrophage colony-stimulating activity
DF	differentiation factor
	pluripoietin
pCSF	pluripotent CSF
GM-CSF ; CSF-GM	granulocyte-macrophage colony-stimulating factor
CSF-α ; CSF-2	colony-stimulating factor α or 2, respectively
MGI-1GM	macrophage-granulocyte inducer 1GM
Eo-CSF	eosinophil CSF
HCGF	hematopoietic cell growth factor
KTGF	keratinocyte-derived T-cell growth factor
NIF-T	T cell-derived neutrophil migration inhibition factor
M-CSF ; CSF-M	macrophage colony-stimulating factor
CSF-1	colony-stimulating factor 1
urinary CSF	
MGI-1M	macrophage-granulocyte inducer 1M
MGF	macrophage growth factor
IL-3	interleukin 3
CSF-2α ; CSF-2β	colony-stimulating factor 2α or 2β, respectively
ECSF	erythroid CSF
EoCSF	eosinophil CSF
MEG-CSF	megakaryocyte CSF
MCSA ; multi CSF	multi-colony-stimulating activity or factor, resp.
Multi HGF	multilineage hemopoietic growth factor
	mixed CSF
CFU-S	colony-forming unit spleen
CFU-SA	colony-forming unit stimulating activity
CFU-S MF	colony-forming unit spleen maintenance factor
HCGF ; HPGF	hematopoietic cell growth factor
MCGF ; MGF	mast cell growth factor
SAF	stem cell activating factor
	histamine producing cell stimulating factor
HP-2	helper peak 2 or hemopoietin 2

Table 2 (continued)

PSF ; PCSA	persisting cell-stimulating factor or activity, respectively
PSH	panspecific hematopoietin
	20α-dehydrogenase-inducing factor
	Thy-1 inducing factor
WGF	WEHI-3 growth factor
BP ; BPA	burst-promoting activity

(Ibelgaufts 1992).

Table 3. Bioassays for the different CSF

CSF	biological basis of assay	cell/assay system
G-CSF	proliferation	NFS60 cell line (human leukemia), GNFS-60 (murine leukemia)
	granulocyte colony formation	bone marrow colony assay
	enhance superoxide production	neutrophils
GM-CSF	proliferation	TF-1 (human erythro-leukemia), MO7e (human megkaryo-blastic leukemia), AML-193 (acute myeloid leukemia)
	granulocyte macrophage colony formation	bone marrow colony assay
	enhance oxidative burst	neutrophils
	degranulation	neutrophils
M-CSF	proliferation	M-NFS60 (murine leukemia)
	macrophage colony formation	bone marrow colony assay
	induce production of IFN, TNF, G-CSF, IL-1	macrophages
IL-3	proliferation	TF-1 (human erythro-leukemia), MO7e (human megkaryo-blastic leukemia), AML-193 (acute myeloid leukemia)
	erythroid, granulocyte and macrophage colony formation	bone marrow colony assay

(Thorpe et al. 1992).

I
G-CSF

1
General Information

1.1
Molecular Biology and Endogenous Production

A
The Molecule G-CSF

The various names originally given to G-CSF (see Table 2) reflect its numerous bioactivities. Only later were these activities traced back to one and the same molecule. Human G-CSF was first purified from the bladder carcinoma cell line 5637 in 1985 (Moore 1991) and subsequently cloned and expressed in *E. coli* (Souza et al. 1986) and monkey COS cells (Tsuchiya et al. 1986; Nagata et al. 1986).

The single copy of the gene encoding G-CSF is situated on the q arm of chromosome 17 near other genes involved in the development of neutrophilic granulocytes (Gabrilove and Jakubowski 1990). A chain of 174 amino acids with a molecular mass of 18 kD forms the protein, which is O-glycosylated in its native form. Therefore, the apparent relative weight is approximately 19.6 kD (Moore 1991). Murine and human G-CSF share 73% homology at the amino acid level, which explains the significant species cross-reactivity (Moore 1991). Structural homology between G-CSF and IL-6 indicates that these two factors may share a common ancestral gene (Ogawa 1993).

Approved pharmaceutical forms of G-CSF include a recombinant non-glycosylated protein expressed in *E. coli* (Filgrastim) and glycosylated forms expressed in Chinese hamster ovary cells or yeast (Lenograstim). Both forms share practically the same pharmacodynamic properties (Yuo et al. 1990), though their pharmacokinetics differ (Nissen et al. 1994). Glycosylation stabilizes the molecule in vitro by suppressing polymerization and/or conformational changes (Oh-eda et al. 1990) and lends resistance to protease degradation in human serum (Nissen et al. 1994). Whether the differences between the glycosylated and non-glycosylated forms of G-CSF are of significance in clinical use has not yet been established (Frampton et al. 1995).

B
Endogenous Production of G-CSF

CSF are produced continuously in some tissues such as in bone marrow. They are also produced readily at the focus of infectious or inflammatory disorders. The endogenous production of the CSF is often overlooked when treating patients with the recombinant material. Therapeutic efficacy can only be reached if there is an absolute or relative CSF-deficiency. Therefore, the endogenous production and its regulation deserve careful attention with regard to putative indications.

Induction of G-CSF

The production of G-CSF is initiated by a number of factors in various cell populations (Table 4).

In resting monocytes, the levels of G-CSF transcripts are low (Ernst et al. 1989). Stimulation of human monocytes with lipopolysaccharide (LPS), peptidoglycan breakdown products, phorbol myristate acetate (PMA) or cycloheximide had little effect on the transcription rate of the gene, but instead resulted in the stabilization of G-CSF mRNA (Vellenga et al. 1988; Ernst et al. 1989; de Wit et al. 1993; Cluitmans et al. 1993; Dokter et al. 1994; Hamilton 1994). Direct subsequent translation of the mRNA to G-CSF protein in these cells was observed in response to LPS (Vellenga et al. 1988; Hamilton et al. 1992b; Sallerfors and Olofsson 1992), phytohemagglutinin (PHA) plus PMA (Oster et al. 1989c), GM-CSF (Oster et al. 1989b; Sallerfors and Olofsson 1992), interleukin-1 (IL-1) (Sallerfors and Olofsson 1992), IL-3 (Oster et al. 1989b), interleukin-4 (IL-4) (Wieser et al. 1989), M-CSF (Ishizaka et al. 1986; Motoyoshi et al. 1989) and peptidoglycan breakdown products (Dokter et al. 1994). Tumor necrosis factor-α (TNF-α) and interferon-γ (IFN-γ) also induced the release of G-CSF by human monocytes separately and synergized in combination in culture (Lu et al. 1988a). Anti-IL-1-antibodies partially blocked G-CSF release by human blood leukocytes in the presence of human serum (Quesniaux et al. 1992), suggesting an auto- or paracrine role for IL-1 in G-CSF induction. Gram-positive stimuli, e.g. heat-killed *Staphylococcus aureus* and lipoteichoic acid (LTA), an element of their cell wall, were potent inducers of G-CSF release in human whole blood (Hartung and Wendel 1998). Further, *Mycobacterium avium* strains have been shown to induce the release of G-CSF in human macrophages in vitro (Fattorini et al. 1994). Mature, alveolar macrophages produced greater amounts of G-CSF in response to LPS in comparison with peripheral blood monocytes (Nelson 1994). Unseparated peripheral blood mononuclear cells

Table 4. Factors affecting the production of the G-CSF protein

Stimulus type	Stimulus	Cell type	Results	References
Bacterial molecules	LPS	monocytes	+	Vellenga et al. 1988
		macrophages	+	Nelson 1994
		PMN	+	Ichinose et al. 1990
		stromal cells	+	Nelson 1994
		mesothelium	+	Demetri et al. 1989
	LTA	whole blood	+	own observations
	Mycobacterium avium	macrophages	+	Fattorini et al. 1994
	heat-killed *S. aureus*	whole blood	+	own observations
	peptidoglycan derivatives	monocytes	+	Dokter et al. 1994
Pro-inflammatory mediators	IL-1	monocytes	+	Sallerfors and Olofsson 1992
		stromal cells	+	Fibbe et al. 1988
		endothelium	+	Seelentag et al. 1987
		fibroblasts	+	Seelentag et al. 1989
		T lymphocytes	+	Lu et al. 1988b
		mesothelium	+	Demetri et al. 1989
	TNF-α	monocytes	+	Lu et al. 1988a
		endothelium	+	Seelentag et al. 1987
		fibroblasts	+	Koeffler et al. 1987
		mesothelium	+	Demetri et al. 1989
	IFN-γ	monocytes	+	Lu et al. 1988a
		T lymphocytes	+	Lu et al. 1988b
Colony stimulating factors	GM-CSF	monocytes	+	Oster et al. 1989c
		PMN	+	Lindemann et al. 1989b
	M-CSF	monocytes	+	Ishizaka et al. 1986
	IL-3	monocytes	+	Oster et al. 1989b
Other	IL-4	monocytes	+	Wieser et al. 1989
	EGF	mesothelium	+	Demetri et al. 1989
	PMA + PHA	monocytes	+	Oster et al. 1989c
		PBMC	+	Oster et al. 1989c

Key: + induction of G-CSF production and secretion.

(PBMC) were noted to release G-CSF in response to PHA plus PMA (Oster et al. 1989c).

Although monocytes and macrophages are considered the major producers of G-CSF, cells which are likely to be near the site of an inflammation, such as endothelial cells, or which are recruited there fastest, i.e. polymorphonuclear granulocytes (PMN), are able to release substantial amounts of the factor when activated by the relevant mediators: LPS (Ichinose et al. 1990) or GM-CSF (Lindemann et al. 1989b) induced a significant G-CSF release from PMN in vitro. Human T-cells responded to TNF-α or IFN-γ with the production of G-CSF; their effects were potentiated when both factors acted in synergy (Lu et al. 1988b).

Human bone marrow stromal cells were induced by IL-1 (Fibbe et al. 1988) or LPS (Nelson et al. 1994) to release G-CSF; umbilical vein endothelial cells responded to IL-1 and TNF with the production of G-CSF (Seelentag et al. 1987; Zsebo et al. 1988). Formation of G-CSF mRNA was described in fibroblasts stimulated with either TNF-α, phorbol ester, cycloheximide (Koeffler et al. 1988) or IL-1(Nelson et al. 1994), but the subsequent release of the G-CSF protein was only described in response to TNF (Koeffler et al. 1987 and 1988; Seelentag et al. 1989; Furman et al. 1992; Shannon et al. 1992) and IL-1 (Seelentag et al. 1989; Shannon et al. 1992; Nelson et al. 1994). Epidermal growth factor (EGF), LPS, TNF-α and IL-1 each induced G-CSF transcripts in human mesothelial cells (Demetri et al. 1989; Lanfrancone et al. 1992).

Furthermore, a number of tumor cell lines, e.g. derived from squamous cell carcinoma and hepatoma (Demetri and Griffin 1991; Lai and Bauman 1996), produced G-CSF. A case of thyroid cancer and its metastatic lesions producing both G-CSF and GM-CSF autonomously has been reported (Nakada et al. 1996).

Modulation of G-CSF production

Apart from the primary mediators which induce the production of G-CSF, other factors may regulate the extent of its production by working with the inductor to cause an additive or a synergistic effect or by canceling out its action, as displayed in Table 5.

LPS- or IL-1-inducible G-CSF production by human monocytes was increased by IFN-γ (Hamilton et al. 1992b; de Wit et al. 1993), prostaglandin E_2 (PGE_2) (Hamilton 1994; Lee et al. 1990) and calcium ionophore A23187 (Vellenga et al. 1991; de Wit et al. 1993), but suppressed by dexamethasone (Hamilton et al. 1992b; Hamilton 1994), indomethacin (Lee et al. 1990; Hamilton 1994) and by IL-4 (Vellenga et al. 1991; Hamilton et al. 1992b; de Wit et al. 1993; Hamilton 1994), a contradiction to the aforementioned in-

Table 5. Modulation of G-CSF production

Cells	Primary stimulus	Augmenting factors	References	Inhibiting factors	References
monocytes/ macrophages	LPS; IL-1	IFN-γ PGE, ionophore A23187	Hamilton et al. 1992b Lee et al. 1990 Vellenga et al. 1991	dexamethasone indomethacin IL-4	Hamilton et al.1992b Lee et al. 1990 Vellenga et al. 1991
	TNF-α; IFN-γ			IL-10 HIV infection	Kruger et al. 1996b Esser et al. 1996
endothelium	IL-1	corticosteroids	Lenhoff and Olofsson 1996	cephalosporins	Lenhoff and Oloffson 1996
mesothelium	TNF; LPS	EGF	Demetri et al. 1989		
stromal cells	IL-1			HIV infection	Moses et al. 1996
fibroblasts	IL-1	IL-4 basic fibroblast growth factor	Hamilton et al. 1992a	IFN-γ dexamethasone cyclooxygenase-inhibitor	Hamilton et al. 1992a
whole blood	LPS	GM-CSF	Hartung and Wendel 1998	IL-10 IL-6 IL-13 IFN-γ	Kruger et al. 1996b Hartung and Wen-del 1998

ductive effects of IL-4. IL-10 inhibited the production of G-CSF at the transcriptional level in monocytes activated with TNF-α and/or IFN-γ (Kruger et al. 1996b). Anti-IL-10-antibodies increased G-CSF release in LPS-stimulated monocytes (Kruger et al. 1996b). When human whole blood was incubated in the presence of LPS, we found a significant inhibition (about 50%) of primarily monocyte G-CSF-formation by IL-6, IL-13 and IFN-γ. In the same series of experiments, IL-1, IL-3, IL-8, IL-10, IL-11 and TNF had no major effect and GM-CSF slightly increased G-CSF release (Hartung and Wendel 1998).

Monocytes and macrophages infected with human immunodeficiency virus (HIV) downregulated G-CSF production, an effect which may provide part of the explanation of neutrophilic dysfunction in HIV-infected patients (Esser et al. 1996). Neither cephalosporins nor corticosteroids affected LPS-stimulated monocyte secretion of G-CSF (Lenhoff and Olofsson 1996). However, in IL-1α-stimulated endothelial cells, the former downregulated and the latter enhanced the production of the factor (Lenhoff and Olofsson 1996).

The combination of EGF with either TNF or LPS induced threefold more G-CSF transcripts in mesothelial cells than did either factor alone (Demetri et al. 1989). Although HIV-infected stromal cultures constitutively expressed normal levels of G-CSF, IL-1α-induced release of G-CSF was reduced (Moses et al. 1996). In this way, the capacity of hematopoietic stroma to respond to regulatory signals, that normally augment blood cell production during periods of increased demand, is reduced (Moses et al. 1996). The stimulatory action of IL-1 on the production of G-CSF by fibroblasts was suppressed by IFN-γ, dexamethasone and cyclooxygenase inhibitors, but potentiated by basic fibroblast growth factor and IL-4 (Hamilton et al. 1992a).

Leukemic cells from some AML (acute myelogenous leukemia) patients were found to secrete cytokines such as TNF and IL-1 to stimulate accessory cells in a paracrine manner to produce G-CSF (Oster et al. 1989a).

Serum G-CSF

G-CSF was detectable in the healthy rat lung, but not in the serum. After challenge with *Pseudomonas aeruginosa*, the amount of G-CSF in the lung rapidly increased and serum levels became measurable by 24 h (Nelson et al. 1994), suggesting an important physiological role for G-CSF in host defense against pneumonia. When mice were challenged with *Listeria monocytogenes*, their endogenous G-CSF serum levels rose to a relative maximum after 48 h, preceding the increase in colony-forming cells in the bone marrow and the subsequent reduction in numbers of viable bacteria, and were still above the normal level after five days (Cheers et al. 1988). The G-CSF serum levels of dogs injected with endotoxin (LPS) increased within two hours, peaked at four hours and had not returned to normal 24 h after challenge (Dale et al. 1992). In a murine fecal peritonitis model (Barsig et al. 1996) and in murine endotoxic shock (Hartung and Wendel 1998), we found significant serum G-CSF levels starting two hours after challenge and reaching a plateau after ten hours at levels of 200 ng/ml.

G-CSF serum levels are only seldom detectable in healthy humans (Watari et al. 1989; Kawakami et al. 1990), but the factor is extractable from all major organs at levels higher than those present in the circulation; tissue production levels can be elevated rapidly (in minutes or hours) after stimulation (Metcalf 1987). Administration of TNF-α to cancer patients resulted in the accumulation of G-CSF in the serum (Furman et al. 1992). After injection of 10 µg/kg G-CSF subcutaneously (s.c.) into humans, serum G-CSF levels remained above 10 ng/ml for 10 to 16 h (Lieschke and Burgess 1992b). Pharmacological doses of G-CSF raised the serum levels of the factor to those attained by endogenous production during infection. Therefore, the potential of G-CSF seems to lie in maintaining the defense signal for a longer time or in initiating the signal earlier by substituting insufficient endogenous production.

In hyperthermic patients with or without neutropenia G-CSF serum levels were elevated compared with those of afebrile control patients (Cebon et al. 1994; Robins et al. 1995), especially in the acute phase of infection (Kawakami et al. 1990). During this period, patients suffering their first bout of infection had significantly higher G-CSF serum levels than patients who had already experienced previous infections (Kawakami et al. 1992).

The level of G-CSF in the serum peaked within three hours of injury in trauma patients and rapidly decreased again during the following seven days, whereas patients with sepsis retained high plasma levels of the factor throughout the duration of the illness, though these decreased significantly during this time in the survivors (Tanaka et al. 1996).

G-CSF serum levels were found to be elevated in a variety of neutropenic disorders: Half of the patients with acute leukemia during induction chemotherapy (Sallerfors and Olofsson 1991) and the same proportion of patients with myelodysplastic syndrome (Watari et al. 1989) displayed increased G-CSF levels, as did five of six patients with aplastic anemia (Omori et al. 1992). For the latter disease, a reverse correlation between blood neutrophil count and serum G-CSF has been demonstrated (Watari et al. 1989). In a patient with cyclic neutropenia, the elevated levels of G-CSF rapidly decreased below detection limit when the PMN count recovered (Misago et al. 1991). The same was observed in children and adults after allogeneic or autologous bone marrow transplantation (Cairo et al. 1992a). Here, the rate of increase or decrease of endogenous G-CSF was thought to be predicative of either failure to engraft or duration of neutropenia (Cairo et al. 1992a). In a patient with autoimmune neutropenia, peripheral PMN counts changed almost in parallel with G-CSF levels (Omori et al. 1992). It is still unclear whether G-CSF is released in direct response to neutropenia or whether its release is only the result of interactions resulting from infections.

It seems that increased neutrophil levels correlate with increased clearance of G-CSF: patients receiving a continuous infusion of G-CSF after melphalan apparently induced additional clearance mechanisms during the second phase of the biphasic neutrophil response. Patients receiving G-CSF after chemotherapy and autologous bone marrow transplantation retained high serum levels while they were still neutropenic (Layton et al. 1989). The mechanism by which exogenous G-CSF is metabolized and excreted is unknown (Frampton et al. 1995).

Notably, the production of G-CSF by leukocytes of neonates was significantly reduced (Cairo 1993). G-CSF levels were significantly higher in preterm newborns compared with full-term newborns and adults (Cairo et al. 1993). A reduced capacity to produce G-CSF has also been shown in HIV patients (D. Pitrak, personal communication). For other patient groups with reduced neutrophil response to infection, e.g. diabetics, no studies on G-CSF production capacity are available yet. Such studies are mandatory to define clinical situations of G-CSF deficiency giving a rationale for therapeutic substitution.

Deficient G-CSF

G-CSF knock-out mice exhibited chronic neutropenia and impaired ability to control infection with *Listeria monocytogenes* (Lieschke et al. 1994a). Their infection-induced granulopoiesis was severely impaired. Similarly, G-CSF-receptor deficient mice had decreased numbers of phenotypically normal neutrophils, decreased numbers of hematopoietic progenitors in the bone marrow and the expansion and terminal differentiation of these progenitors to granulocytes was impaired (Liu et al. 1996). Furthermore, isolated neutrophils from these mice had an increased rate of apoptosis.

When rats were passively immunized with rabbit anti-G-CSF-antibodies, intrabronchial application of *Pseudomonas aeruginosa* induced diminished PMN recruitment and bactericidal activity (Nelson 1994), but the circulating PMN counts remained normal. In contrast, when we injected mice with sheep anti-murine-G-CSF-antiserum they developed severe neutropenia. These animals became much more susceptible to infection with a human stool suspension and died of otherwise sublethal fecal peritonitis. If, on the other hand, the anti-G-CSF-antiserum was injected at the time of infection, it had no effect on survival or early neutrophilia (3 h) but blunted the late neutrophilia (12 h) (Barsig et al. 1996). Although human G-CSF did not elicit antibodies in humans, dogs given human G-CSF repeatedly developed autoimmune neutropenia due to the formation of autoantibodies which neutralized the endogenous canine G-CSF (Hammond et al. 1991).

Patients with severe congenital neutropenia (SCN; Kostmann Syndrome) have increased G-CSF serum levels, yet myelopoieis is arrested at promyelocyte stage and neutrophils are absent in bone marrow and blood. The response of their neutrophil precursors to endogenous G-CSF is defective (Mempel et al. 1991). This may be explained by a nonsense mutation of the G-CSF receptor gene found in five patients with the disease (Dong et al. 1994 and 1997). When this mutation was expressed in murine myeloid cells, it transduced a strong growth signal but was defective in maturation induction (Dong et al. 1994).

These findings indicate that G-CSF is a major player in the regulation of granulopoiesis, both in normal and emergency situations, but that there also exist alternative or additional regulatory mechanisms for neutrophil production.

Excess G-CSF

Lethally irradiated mice transplanted with marrow cells expressing G-CSF through a retroviral vector had high serum G-CSF levels and mostly remained healthy. Neutrophilic granulocytosis and tissue infiltration of their lung and liver was observed and spleen, peritoneal and peripheral blood cellularity and progenitor numbers increased, but total bone marrow cell counts remained unaffected. No tumors developed regardless of chronic G-CSF stimulation (Chang et al. 1989).

Toxicity

The most common side effects of short-term G-CSF use are bone pain and myalgia, headache and tiredness (Schwab and Hecht 1994). These are usually mild and transient and can be treated with analgesics (Frank and Mandell 1995). Local inflammation at sites of injections and asymptomatic elevations of lactate dehydrogenase and alkaline phosphatase have also been observed. Splenomegaly was seen in one third of children who received G-CSF for chronic neutropenia, but became symptomatic in only 10% (Frank and Mandell 1995). Considering that more than 2 million patients have been treated with G-CSF already, it is remarkable that reports of more severe side-effects are so rare.

C
Receptors and Signal Transduction

The gene for the G-CSF receptor (G-CSF-R, CD114) is located on the short arm of chromosome 1. Three distinct human G-CSF receptor cDNA have been obtained. The major human G-CSF receptor has a calculated molecular weight of 90 kD, but an apparent molecular mass of 150 kD due to extensive glycosylation. It is an 813 amino acid polypeptide with a single transmembrane domain, a small cytoplasmic domain and a large extracellular domain with the tryptophane-serine-x-tryptophane-serine (WSXWS) motif (Rapoport 1992). One of the three cDNA isolated from human cells lacked most of the transmembrane sequence, suggesting that it encodes a soluble, secreted form of the receptor. The physiological role of such a receptor is unclear; it may serve as a 'sink' for eliminating excess cytokines and thus modulate their activities. The mouse and human G-CSF receptor cDNA share more than 60% homology at the amino acid level, consistent with the ability of both human and murine G-CSF to bind to both species' G-CSF receptors (Rapoport 1992).

PMN displayed about 300 to 1000 G-CSF high-affinity receptors per cell, depending on their level of differentiation, which can already effect biological responses at low levels of occupancy (Moore 1991; Demetri and Griffin 1991; Nelson 1994). A low affinity binding activity has been described in the murine myeloid NFS-60 cell line, which is probably derived from the monomeric G-CSF-R protein, whereas high affinity is the result of homodimerization (Avalos 1996). When G-CSF binds to its receptor, the complex is internalized, a mechanism necessary for G-CSF to perform its functional activity (Schwab and Hecht 1994). The G-CSF receptors were rapidly downmodulated when human neutrophils were incubated with the neutrophil-activating agents LPS, fMLP (N-formylmethionyl-leucyl-phenylalanine) or GM-CSF (Nicola et al. 1986; Rapoport 1992).

Myeloid progenitor cells, monocytes, macrophages, platelets, some T and B lymphoid cell lines, endothelial cells, placenta and trophoblastic cells also all express G-CSF receptors, but eosinophils and erythroid cells do not (Nicola and Metcalf 1985; Shimoda et al. 1990 and 1993; Moore 1991; Avalos 1996). Acute nonlymphocytic leukemia cells classified as M4 and the murine WEHI-3b(D^+) cells display G-CSF receptors and were induced by G-CSF to undergo terminal differentiation to macrophages and granulocytes (Souza et al. 1986). Furthermore, the myeloid leukemia cell lines KG-1 (Avalos et al. 1990) and HL-60 (Welte et al. 1987; Avalos et al. 1990) as well as the small-cell lung carcinoma cell lines H128 and H69 have high affinity binding sites for G-CSF (Avalos et al. 1990). TNF-α reduced G-CSF-R number on murine peritoneal exsudate macrophages transiently in a time- and dose-dependent process (Shieh et al. 1991). GM-CSF, M-CSF and G-CSF all reduced the G-CSF binding capacity of murine macrophages, but IL-1 and IFN-γ had no such effect (Avalos et al. 1990; Shieh et al. 1991).

When a plasmid for G-CSF receptor expression was introduced into a line of mouse myeloid precursor cells that are normally unresponsive to G-CSF, the proliferation of these cells was stimulated by G-CSF and expression of neutrophil-specific genes such as myeloperoxidase (MPO) and leukocyte elastase was induced (Fukunaga et al. 1993). In this model, GM-CSF and IL-3 inhibited G-CSF receptor-mediated MPO gene expression. Mutational analysis of the receptor identified a region of the cytoplasmic domain which is sufficient to transmit the proliferation signal into cells, while another region plays an essential role in transducing the differentiation signal.

The precise second messenger pathways used by G-CSF have not yet been identified. G-CSF had no direct effect on membrane depolarization, Ca^{2+} flux or intracellular free Ca^{2+} concentrations (Rapoport 1992). Prior to the proliferative response induced by G-CSF binding to murine myeloblastic NFS-60 cells, a time-dependent activation of the guanosine triphosphate-binding proteins and the adenylate cyclase system has been observed (Matsuda et al. 1989).

1.2
Role in Hematopoiesis

Studies using clonal transfer and the delayed addition of other regulators showed that G-CSF could directly stimulate the initial proliferation of a large proportion of the granulocyte-macrophage progenitors in adult bone marrow and also the survival and/or proliferation of some multipotential, erythroid and eosinophil progenitors in fetal mouse liver. However, G-CSF alone was unable to sustain continued proliferation of these cells resulting in colony-formation (Metcalf and Nicola 1983). But, G-CSF may act in synergy with IL-3 to support lineages, such as megakaryocytes, resulting in platelet production (Gabrilove and Jakubowski 1990) as well as early pluripotential stem cells, accelerating their entry into cell cycle (Moore 1991) and pre-B-cells, promoting their activation and growth (Gabrilove and Jakubowski 1990). Low concentrations of G-CSF predominantly stimulated pure PMN colonies, while high concentrations of G-CSF resulted in the development of macrophage-containing colonies (Metcalf and Nicola 1983; Moore 1991). TNF-α, TNF-β and IFN-γ all had a suppressive effect on human cell colonies formed after stimulation with G-CSF in vitro (Barber et al. 1987). This action of IFN-γ was antagonized by IL-4 (Snoeck et al. 1993). G-CSF in vitro had no effect on murine T-cell proliferation (Aoki et al. 1995) or on the mixed lymphocyte reaction of human blood mononuclear cells in vitro (Kitabayashi et al. 1995).

In experiments with rodents, injection of G-CSF resulted in neutrophilia by increasing the production of PMN in the bone marrow and hastening the release of less mature forms into the circulation (Tamura et al. 1987; Ulich et al. 1989; Cairo et al. 1990a; Tkatch and Tweardy 1993). G-CSF was able to cross the placenta, stimulate fetal rat granulopoiesis and to augment the marrow and spleen neutrophil-storage pools (Medlock et al. 1993). Repetitive s.c. injection of G-CSF for seven days into rats resulted in a biphasic increase in PMN counts (Ulich et al. 1989). In mice (Tamura et al. 1987) or humans, no such transient negative-feedback mechanism on PMN release from bone marrow was found. The increase in monocyte production in mice receiving G-CSF (Lord et al. 1991) seems to be attributable to its stimulation of endogenous M-CSF production, as it could be selectively canceled out by M-CSF antiserum (Gilmore et al. 1995). G-CSF obliterated the cycling of neutrophils, platelets and reticulocytes in dogs with cyclic neutropenia (Hammond et al. 1990).

In vivo experiments with cynomolgus monkeys treated with s.c. doses of G-CSF for 14 to 28 d resulted in dose-dependent increases in the peripheral white blood cell count (WBC), which reached a plateau after 1 week of G-CSF treatment. The elevation of WBC was the result of an increase in the absolute neutrophil count (ANC) (Welte et al. 1987). In cyclophosphamide-induced myelosuppression, G-CSF shortened the time of WBC recovery in two treated monkeys to one week, as compared with more than four weeks in the control monkey (Welte et al. 1987).

Thirty minutes after healthy volunteers were injected with G-CSF s.c., their circulating PMN counts declined, probably owing to increased adherence of the activated cells to the endothelium of blood vessels of the lung (Morstyn et al. 1988; Bronchud et al. 1988; Katoh et al. 1992). We made similar observations in healthy volunteers, where the PMN count fell below 500 PMN/μl 30 min after s.c. injection of G-CSF. The subsequent increase in PMN is attributable to the release of less mature PMN from the bone marrow, not to recruitment from this marginated pool (Katoh et al. 1992). Doses of G-CSF ranging from 1 to 60 μg/kg/d given for five or six days produced dose-dependent increases of 1.8- to 12-fold of the original ANC (Dale 1994). Blood neutrophil half-life was not altered significantly by G-CSF treatment. In a recent human volunteer study, we found a dose-dependent increased PMN count with a plateau lasting for the whole treatment with G-CSF. 72 h after the last injection, counts equal to those prior to treatment were determined (Hartung and Wendel 1998). No significant differences in PMN recruitment were observed when comparing the parameters of young and elderly subjects (Dale 1994; Price et al. 1996). Dexamethasone increased the levels of neutrophilia induced in normal subjects by G-CSF, which indicates that such a combination may be useful for the mobilization of PMN in the peripheral blood of granulocyte donors (Liles et al. 1997b).

Human volunteers injected with 480 μg of G-CSF s.c. developed monocytosis, i.e. an approximately 3-fold increase of blood monocytes was observed within one day (Hartung et al. 1995a). In a volunteer study, we also followed an increase in monocyte numbers until day 8 of daily 300 μg G-CSF administration, which was maximally 5-fold, after which the count fell despite continued treatment (submitted for publication). A different study recorded no change in monocyte counts and an increased proportion of immature myeloid cells as well as an increase in the levels of both erythroid and myeloid precursor cells in the bone marrow (Asano et al. 1988). G-CSF infusion [30 μg/kg] in non-neutropenic patients with advanced malignancies resulted in a transient but nearly complete decrease of monocytes after 15 to 30 min, which normalized after 60 min (Lindemann et al. 1989a). At later time points and high doses of G-CSF [30–60 μg/kg] an up to tenfold monocytosis was observed (Morstyn et al. 1988; Gabrilove et al. 1988). Smaller G-CSF doses resulted in a threefold increase in monocyte counts (van der Wouw et al. 1991), which correlated with an increase in CFU-GM, i.e. granulocyte and monocyte precursors, in bone marrow (Liu et al. 1993).

A phase I/II trial with 30 cancer patients receiving G-CSF documented a dose-related increase in the absolute number of circulating progenitor cells of granulocyte-macrophage, erythroid, and megakaryocyte lineages, which remained elevated two days after the cessation of therapy. Here, the relative frequency of different types of progenitor cells in peripheral blood remained unchanged. Also, the frequency of progenitor cells in the marrow was variable after G-CSF treatment, but in most patients was slightly decreased (Dührsen et al. 1988).

Effects of G-CSF treatment on lymphocyte counts in humans were already noted in the very early reports (Morstyn et al. 1988; Kerrigan et al. 1989). Treatment with high doses of G-CSF [10 or 100 μg/kg/d] was associated with a 1.5- to 2.5-fold increase in absolute lymphocyte counts (Demetri and Griffin 1991). A relative lymphocytosis was also documented in a study carried out in our laboratory with volunteers treated repeatedly with G-CSF resulting in a maximal lymphocyte count on day 8. Lymphocyte subtyping (CD4, CD8, CD16, i.e. natural killer cells and CD19, i.e. B cells) revealed that all subpopulations increased during treatment, without major changes in the proportions. Furthermore, we observed an augmented proliferation of lymphocytes in response to anti-CD3-antibodies or PHA, or through initiation of a mixed lymphocyte reaction with Daudi cells ex vivo. However, at times later than 72 h after initiation of treatment, the proliferative response of lymphocytes was markedly suppressed. During the twelve days of treatment, there was no effect on NK (natural killer) cell function (submitted for publication). Another study on nine healthy volunteers reported an increase in absolute lymphocyte and monocyte counts after four days of G-CSF administration (Sica et al. 1996). Of the lymphocyte subsets analyzed, $CD3^+$, $CD19^+$, NK cells and activated T-lymphocytes collected in the leukapheresis product were all increased (Sica et al. 1996). G-CSF aided in the recovery of $CD8^+$ T-cells after autologous bone marrow transplantation; this $CD8^+$ regeneration was produced mainly by activated cells ($CD38^+$/HLA-DR^+) lacking CD11b antigen (Miguel et al. 1996).

G-CSF further enhanced colony-formation by the human myeloid leukemic cell lines HL-60 and KG-1 as well as by non-hematopoietic small cell lung cancer lines H128 and H69 (Avalos et al. 1990).

Risk assessment studies found no or only minimal suppression of erythropoiesis and no suppression of the production of other cell types in short- and long-term studies, but explained the shift in relative proportions of cell types with an unchanged population size of erythroblasts and a concomitant overwhelming increase of granulocytes (Keller and Smalling 1993).

2 Effects on Granulocytes

2.1 Effects on the Functions of Neutrophilic Granulocytes

Neutrophilic granulocytes have a variety of defense mechanisms against invading organisms: ingestion of the particles or cells; release of reactive oxygen species (ROS), i.e. oxidative burst; degranulation of catabolic enzymes and production of mediators to coordinate their activity with other cells involved in the immune response. As these weapons can in part be detrimental to the body's own cells (Smith 1994), they must be employed only at the site of invasion.

The pleiotropic nature of G-CSF is already apparent when its actions on neutrophils only are considered. Over all, G-CSF appears to induce only minor PMN responses by itself, but rather modifies the response of these cells to subsequent stimulation: While phagocytosis and oxidative burst are augmented, G-CSF barely affects degranulation and chemotaxis. This is the benefit of the priming effect G-CSF has on PMN: their functions are potentiated in case of stimulation by exogenous signals. On the one hand, influences of G-CSF on surface molecule expression were attributed at least in part to the recruitment of PMN in different stages of maturation. On the other hand, there was a de novo expression of CD64 associated with increased antibody-dependent cellular toxicity (Valerius et al. 1993). Mediator release by PMN is shifted by G-CSF in favor of an anti-inflammatory balance, i.e. enhanced release of IL-1ra and IFN-α and decreased formation of the chemotactic leukotriene B_4.

A
Phagocytosis

Pure G-CSF had no bactericidal effect as such (Souza 1990). Phagocytosis by neutrophils was greatly augmented by opsonization of particles with either antibodies or complement. G-CSF induced a rapid change from low- to high-affinity neutrophil CD89 (IgA Fc receptors), which was associated with IgA-mediated phagocytosis (Weisbart et al. 1988). A number of studies have been performed on the phagocytosis and killing of microbes. PMN exposed in vitro to G-CSF acquired an increased potency to kill opsonized *E. coli* (Kropec et al. 1995; Daschner et al. 1995; McKenna et al. 1996) and *Staphylococcus aureus* (Roilides et al. 1991; Bober et al. 1995a; McKenna et al. 1996). Increased *Micrococcus lysodeikticus* and yeast particle phagocytosis by PMN in vitro has also been observed (Bober et al. 1995a). Variable results were reported as to the in vitro efficacy of G-CSF in increasing *Candida albicans* killing (Roilides et al. 1991; Yamamoto et al. 1993; Roilides et al. 1995; Bober et al. 1995a). In one of these studies, PMN from healthy volunteers incubated with G-CSF displayed enhanced antifungal activity toward *Candida tropicalis* and *Candida albicans,* but toward *Candida parapsilosis* only when incubated with high concentrations of G-CSF (Roilides et al. 1995).

Phagocytotic activity of G-CSF-treated hamster neutrophils was determined with opsonized latex beads in vitro, but was not significantly stimulated ex vivo (Cohen et al. 1988). By challenging whole blood with *Salmonella abortus equi* CFU, we found that the bactericidal activity of blood from G-CSF-treated healthy volunteers was increased by several orders of magnitude in comparison to blood from a placebo-treated control group. The augmented killing activity ex vivo was shown to be mediated by PMN (unpublished observations).

B
Oxidative Burst

G-CSF itself failed to induce an oxidative burst by isolated human PMN in suspension (Nathan 1989). G-CSF did not induce superoxide release in neutrophils adherent to fetal bovine serum-coated or uncoated plates (Kadota et al. 1990). But, PMN adherent to laminin or membrane proteins were reported to react to G-CSF directly with a massive respiratory burst similar in extent to that obtained by stimulation with TNF-α and TNF-β (Nathan 1989).

In vitro, G-CSF primed PMN for an enhanced oxidative burst triggered by the bacterial chemotactic peptide fMLP (Kitagawa et al. 1987; Nathan 1989; Kadota et al. 1990; Yuo et al. 1990; Balazovich et al. 1991; Khwaja et al. 1992; Roilides et al. 1992; Yamamoto et al. 1993; Sullivan et al. 1993), ionomycin (Balazovich et al. 1991), wheat germ agglutinin (Yuo et al. 1989), complement factor C5a (Khwaja et al. 1992), opsonized zymosan (Katsura et al. 1993) and *E. coli* α-hemolysin (Konig and Konig 1994). However, there are also reports indicating that G-CSF has no effect on the release of ROS mediated by fMLP (Treweeke et al. 1994), concanavalin A (ConA), ionomycin or phorbol ester (Yuo et al. 1989; Kadota et al. 1990), of which the latter two bypass the receptors to stimulate the cells. G-CSF enhanced PMN oxidative burst in response to opsonized blastoconidia and pseudohyphae of *Candida albicans* (Roilides et al. 1992) and toward *Aspergillus fumigatus* hyphae (Roilides et al. 1993). Apparently, TNF synergized with G-CSF in priming the cells for enhanced oxidative burst (Khwaja et al. 1992), while protein kinase C inhibitors or pertussis toxin inhibited the actions of G-CSF in this respect (Balazovich et al. 1991). The impaired oxidative burst in response to fMLP of PMN from rats made diabetic by streptozocin treatment could be partially restored by G-CSF in vitro (Sato and Shimizu 1993).

Neutrophils isolated from G-CSF-treated hamsters displayed an increased oxidative burst in reaction to opsonized zymosan particles (Cohen et al. 1988). G-CSF increased ROS release by PMN in response to opsonized zymosan from neonatal or adult animals ex vivo (Wheeler et al. 1994). Peritoneal exsudate cells from G-CSF-treated rats displayed a greater superoxide production in response to fMLP than controls, which in turn produced more superoxide on stimulation than peripheral blood neutrophils (Murata et al. 1995). After cecal ligation and puncture, rats showed a suppressed ex vivo PMA-inducible burst, which was restored by in vivo G-CSF treatment (Goya et al. 1993).

When PMN were isolated from G-CSF-treated individuals, patients or volunteers, an augmentation of ex vivo fMLP-, ionomycin- and C5a-stimulated burst similar to in vitro results was recorded (Lindemann et al. 1989a; Ohsaka et al. 1989; Balazovich et al. 1991; Khwaja et al. 1992; Weiss et al. 1994 and 1995; Iacobini et al. 1995), though G-CSF appeared to fail to increase burst stimulated by phorbol ester (Ohsaka et al. 1989; Iacobini et al. 1995; own unpublished results) or zymosan (Weiss et al. 1994 and 1995). Furthermore, G-CSF primed neutrophils for sustained respiratory burst in response to extracts of *Candida albicans*, *Aspergillus fumigatus* and *Rhizopus arrhizus* (Liles et al. 1997a). Daily in vivo G-CSF treatment did not affect the oxidase activity per neutrophil in whole blood, though chemical and opsonin-stimulated MPO oxygenation activities per neutrophil were greatly increased (Allen et al. 1997). Furthermore, superoxide production, which was impaired in AIDS patients in comparison to controls, was greatly augmented ex vivo in response to zymosan 48–72 h after s.c. administration of G-CSF (Vecciarelli et al. 1995).

C
Adhesion and Chemotaxis

G-CSF had no effect on the adhesion of neutrophils to endothelium in vitro (Yong 1996). Instead, the adhesion of PMN to nylon fibers was augmented (Yuo et al. 1989 and 1990). Peripheral blood neutrophils of rats treated in vivo with G-CSF showed less adherence to plastic plates coated with fetal calf serum, whether they were stimulated with fMLP or TNF-α or neither, in comparison to PMN from control rats, but peritoneal exsudate neutrophils from the same G-CSF-treated rats adhered more than those of control rats (Murata et al. 1995).

G-CSF itself is weakly chemokinetic, i.e. it promoted migration of neutrophils independent of a gradient (Smith et al. 1994). In vitro, G-CSF was shown to induce a chemotactic response of PMN at fMLP concentrations below the normal minimum required for reaction (Wang et al. 1988; Bober et al. 1995a). Leukotriene B_4 (LTB_4)-inducible chemotaxis was not altered (Bober et al. 1995a). The effect of G-CSF on trans-endothelial migration in vitro across unstimulated endothelium is controversial, but there seems to be a consensus that migration across TNF-α- or IL-1-stimulated endothelium is not affected (Smith et al. 1994; Yong 1996).

In rodents, G-CSF treatment restored experimentally impaired chemotaxis, i.e. after thermal injury in mice (Sartorelli et al. 1991) or after peritonitis brought about by cecal ligation and puncture in rats (Goya et al. 1993). Nevertheless, in the model of s.c. *E. coli* injection into rats, G-CSF pretreatment resulted in less infiltration as assessed by local MPO activity and glucose uptake compared with controls (Lang et al. 1992a), however, at the same time resulted in a better elimination of bacteria. Even in mice with inactivated G-CSF genes, the typical accumulation of neutrophils in response to an injection of casein containing bacteria was observed (Metcalf et al. 1996). These conflicting data suggest that the positive results may have been caused by the rapid recruitment of a greater number of PMN into the circulation rather than by a stimulatory effect of G-CSF on their chemotaxis.

Furthermore, we have explored the in vivo effects of G-CSF alone and G-CSF pretreatment before LPS exposure on the microvasculature of the rat liver by intravital fluorescence microscopy (Vollmar et al. 1997): G-CSF alone enhanced the adhesion of leukocytes to endothelium, but did not affect the activity of the Kupffer cells. In contrast, G-CSF treatment followed by LPS administration attenuated the leukocyte adhesion due to LPS stimulation, reduced the Kupffer cells' activity and protected against microcirculatory perfusion failure and hepatic dysfunction.

Consistent with this interpretation are observations in this respect in human volunteers or patients: In human volunteers, G-CSF treatment resulted in a 50% reduction of PMN influx into implanted skin chambers (Price et al. 1996) and, in another study, in minimal reduction of PMN entry into skin windows (Demetri et al. 1990). G-CSF treatment for local skin abrasion in healthy volunteers did not attract increased numbers of PMN to the site (Dale 1994). No effect of G-CSF infusion on PMN chemotaxis was found in a group of 12 patients with small-cell lung cancer undergoing chemotherapy (Bronchud et al. 1988).

D
Cytotoxicity

The antibody-dependent cellular cytotoxicity (ADCC) of neutrophils from normal mouse bone marrow neutrophils or induced peritoneal neutrophils toward antibody-coated thymoma cells was enhanced in the presence of G-CSF. In this respect G-CSF and GM-CSF showed an additive effect (Lopez et al. 1983), although only G-CSF, not GM-CSF, alone promoted this activity in human neutrophils. G-CSF enhanced human neutrophil-mediated ADCC toward a mastocytoma and two different thymoma target cell lines (Vadas et al. 1983) as well as toward neuroectodermal tumor target cells in vitro, an activity which was inhibited by FcγRII antibodies (Baldwin et al. 1993). G-CSF also augmented the cytotoxic function of human neutrophils toward HIV-infected MOLT-3A cells in vitro (Baldwin et al. 1989). Neutrophils are apparently not infected by HIV, therefore they offer an ideal focus for enhancement of their cytotoxic function in patients with AIDS.

When the tumor-cell growth inhibition by peripheral blood neutrophils and peritoneal exsudate neutrophils toward the RL1 targets from G-CSF-treated rats was compared with control cells from normal animals, it was found that cytotoxic activity of the peritoneal exsudate cells outweighed that of controls, but peripheral blood neutrophils from G-CSF-treated rats were less active in this respect than cells from control animals (Murata et al. 1995).

E
Mediator and Enzyme Synthesis

In vitro, G-CSF was only a weak direct inducer of the secretion of arachidonic acid metabolites (Sullivan et al. 1987; Di Persio et al. 1988c; Herrmann et al. 1990), but it primed neutrophils for enhanced arachidonic acid release in response to the calcium ionophore A23187 (Di Persio et al. 1988c). The priming of neutrophils with G-CSF before exposure to leukocidin from *Staphylococcus aureus* or calcium ionophore resulted in a substantial increase in LTB_4 formation (Hensler et al. 1994). However, G-CSF had no effect in vitro on LTB_4 release by neutrophils challenged with *E. coli* α-hemolysin (Konig and Konig 1994). G-CSF was reported to initiate the release of IFN-α by PMN in vitro (Shirafuji et al. 1990), while it had no effect on IL-8 release by neutrophils in the presence of *E. coli* α-hemolysin (Konig and Konig 1994).

Ex vivo, in neutrophils from G-CSF-treated volunteers, LPS-inducible IL-1 receptor antagonist (IL-1ra) release was increased in whole blood compared with controls, while the shedding of soluble TNF receptors was unaffected when calculated per PMN, which were of course present in higher numbers in the blood of G-CSF-treated donors (Hartung et al. 1995a and 1995b). Concomitantly, their LTB_4 release capacity on stimulation with LPS per individual cell was decreased significantly (unpublished observations).

The IL-1ra, TNF and sTNF-R (soluble TNF receptor) p55 and p75 plasma levels increased in healthy volunteers injected with a single dose of G-CSF. After a second single dose one week later, when all values had returned to baseline, the rise in IL-1ra was greater, the increment in sTNF-R p75 was smaller and the other two increases were similar to the first attempt (Pollmächer et al. 1996). In 14 patients with myelodysplastic syndromes, combined treatment with all-trans retinoic acid and G-CSF increased serum concentrations of soluble TNF receptors (Ganser et al. 1994).

F
Degranulation

The peroxidase-positive (azurophil or primary) granules of human PMN contain MPO, the protease elastase, bacterial permeability increasing protein (BPI) and β-glucuronidase. Markers for peroxidase-negative (specific or secondary) granules are the iron binding protein lactoferrin and neutrophil alkaline phosphatase (NAP) (Borregaard et al. 1993).

In vitro, G-CSF had no effect on fMLP-inducible release of MPO (Treweeke et al. 1994) or on α-hemolysin-induced release of β-glucuronidase (Konig and Konig 1994). However, the activity of NAP was enhanced by G-CSF in vitro in PMN from normal volunteers and to a greater extent in PMN from patients with chronic myelogenous leukemia (Teshima et al. 1990). However, this effect could be suppressed by GM-CSF in a dose-dependent manner when the PMN were incubated with both CSF (Teshima et al. 1990).

In ex vivo experiments carried out with samples from G-CSF-treated healthy volunteers, we found no change of degranulation of primary granules (MPO, elastase, BPI). Contrary to our expectation, we observed a 50% reduction of secondary granule release (lactoferrin) in LPS-stimulated blood from G-CSF-treated healthy volunteers (unpublished observations).

PMN from G-CSF-treated volunteers displayed an increased number of primary granules compared with cells from untreated control donors (Dale 1994). The levels of plasma elastase, bound to its physiologic inhibitor α1-antitrypsin, increased within one hour of injection of G-CSF into healthy volunteers (de Haas et al. 1994). Increased elastase serum levels were reported in non-neutropenic patients with advanced malignancies receiving G-CSF therapy (Lindemann et al. 1989a). The neutrophils of healthy volunteers injected s.c. with G-CSF contained significantly decreased levels of NAP during the first four hours post injection, but thereafter a sharp rise in NAP was measured. Moreover, the plasma levels of lactoferrin increased significantly within one hour (de Haas et al. 1994). Patients with urothelial cancer treated with G-CSF before chemotherapy also had PMN containing increased amounts of NAP (Gabrilove et al. 1988).

Therefore, it appears that G-CSF has only little impact on degranulation in vitro or ex vivo, even though it changes the granule content of PMN in vivo.

G
Expression of Surface Molecules

The mature neutrophil expresses complement receptors CR1 (CD35) and CR3. CR3 receptors are composed of an identical β subunit (CD18) plus a different α subunit (CD11a, CD11b = Mo1, CD11c), corresponding to LFA-1, Mac-1(= C3bi receptor) and P150,95 (Gillan et al. 1993). Fc receptors for immunoglobulin G, including FcγRI (CD64), FcγRII (CD32) and FcγRIII (CD16) are found on mature neutrophils (Gillan et al. 1993). Adhesion is thought to be modulated primarily by interaction between LFA-1 (CD11a/18 complex) on the surface of the neutrophil, with receptors on vessel endothelium (Gillan et al. 1993).

The available data on the effect of G-CSF on the expression of neutrophil surface receptors in vitro are confusing. Apparently, G-CSF has no effect on the expression of CD54 (ICAM-1) (Bober et al. 1995a) or fMLP-receptor (Yuo et al. 1989), but upregulates CD89 (IgA receptor) (Dale et al. 1995) expression in vitro. The CD11b/CD18 expression has been reported as unaffected (Demetri et al. 1990; Treweeke et al. 1994; Bober et al. 1995a) or upregulated (Yuo et al. 1989 and 1990; Linch 1992; Liles et al. 1994a; Dale et al. 1995; Yong 1996), and one study determined that upregulation only took place in neonatal, but not in adult rat granulocytes in vitro (Wheeler et al. 1994). CD35 has also been described as unchanged (Bober et al. 1995a) or upregulated (Dale et al. 1995), whereas CD62L (LAM-1; L-selectin) has been found to be increased (Demetri et al. 1990), decreased (Spertini et al. 1991; Yong and Linch 1992; Ohsaka et al. 1993; Liles et al. 1994a; Dale et al. 1995; Yong 1996) or unaffected (Griffin et al. 1990). The expression of CD64 was only upregulated by G-CSF in bone marrow cultures, not in peripheral blood neutrophils in vitro (Kerst et al. 1993a). The duration of incubation of the neutrophils with G-CSF seems to play a role in their reaction to the cytokine with regard to receptor expression (Liles et al. 1994a).

Some of these great differences in the in vitro results become clearer when the in vivo measurements are considered. G-CSF apparently has different effects on the peripheral blood PMN and on the granulocytes newly recruited from the bone marrow. The former are distinct within 30min after administration of G-CSF parallel to the transient neutropenia. Increased expression of CD11b/CD18 (Ohsaka et al. 1989; Katoh et al. 1992; de Haas et al. 1994), CD62L (Yong and Linch 1992), CD66b (de Haas et al. 1994) and CD16 (Kerst et al. 1993b; de Haas et al. 1994) have been measured, while the values of CD14 receptor expression were not affected (Kerst et al. 1993b). The effects on the bone marrow only come into play later, when the affected granulocytes have entered the peripheral blood stream: Within hours of administration of G-CSF, the CD11b/CD18 expression remained above baseline (Yong and Linch 1992) or returned to baseline (Demetri et al. 1990; de Haas et al. 1994); CD62L returned to baseline values (Demetri et al. 1990; Yong and Linch 1992) as did CD66b (de Haas et al. 1994) and CD16 (Kerst et al. 1993b; de Haas et al. 1994). An increase in the plasma levels of soluble CD16 was measured (de Haas et al. 1994). CD14, an opsonic receptor for LPS-binding proteins was expressed at higher levels on PMN in response to G-CSF (Kerst et al. 1993b; Hansen et al. 1993; Dale et al. 1995), as was CD64, which is usually restricted to mononuclear cells (Repp et al. 1991; Dale et al. 1995) and CD62L (Dale et al. 1995). In another case, no change in the expression of CD62L was found throughout the experiment (de Haas et al. 1994). Neither the expression of CD32 (Kerst et al. 1993b; de Haas et al. 1994) nor CD 63, a marker for primary granules (de Haas et al. 1994), was changed by G-CSF.

After five consecutive days of G-CSF administration, the neutrophils of normal human subjects displayed elevated CD62L and CD14 expression, but decreased expression of CD11b and CD18 in comparison to their levels before the treatment (Liles et al. 1994a). In human healthy volunteers treated with G-CSF for twelve days, we observed shedding of CD62L with concomitantly increased serum levels of soluble L-selectin. Other PMN surface markers such as CD11b/CD18, CD14 and CD71 were hardly affected. However, we also observed the well-known increase in CD64 which was paralleled by decreased CD16 (unpublished results).

Similarly, in 10 postoperative/posttraumatic patients treated with low-dose G-CSF infusions, CD64 expression of PMN was increased by 50% (Weiss et al. 1994). G-CSF also restored the CD64 receptor number and biologic activity mediated by this receptor, i.e. oxidative metabolism and primary granule degranulation, on PMN from patients with septic shock (Simms and D'Amico 1994).

In summary, the available data on PMN surface markers are not completely consistent. Divergent results may be explained by differential stimulation of PMN during isolation procedures. In any case, different treatment regimes and the heterogeneity of the patients' underlying diseases make direct comparisons impossible.

H
Microbial Killing

Ex vivo, G-CSF enhanced PMN-mediated killing of *Aspergillus fumigatus* and *Rhizopus arrhizus* 4-fold and 15-fold, respectively. In contrast, the killing of *Candida albicans* by PMN was unaffected by G-CSF (Liles et al.

1997a). However, some PMN from patients with AIDS showed significantly impaired destructive activity against *Candida albicans* or encapsulated or acapsular *Cryptococcus neoformans*. This could be rectified by s.c. administration of G-CSF (Vecciarelli et al. 1995).

I
Other

G-CSF interfered with the physiological apoptosis of PMN (Williams et al. 1990; Colotta et al. 1992; Adachi et al. 1993; Liles et al. 1994b; Sullivan et al. 1996), apparently by suppressing CD95-mediated apoptosis (Liles et al. 1996) and may thus have importance in regulating the sizes of normal hematopoietic precursor populations in the bone marrow (Williams et al. 1990). Furthermore, local G-CSF production at the focus of infection and inflammation might prolong PMN lifetime by this mechanism. Apart from prolonging neutrophil survival in vitro, G-CSF also preserved the oxidative burst capacity for longer as with the CD16 and CD18 expression and maintained bactericidal function (Ichinose et al. 1990; Sullivan et al. 1996). Protein kinase C inhibitors counteracted this anti-apoptotic effect (Adachi et al. 1993). As this effect of G-CSF on PMN takes place in isolated cells and because it lengthens the survival time at concentrations below those required for colony- stimulating activity, it may be considered a separate activity from bone marrow stimulation (Nelson 1994). Further, the rapid apoptosis of PMN from patients with AIDS could be attenuated in vitro by incubation with G-CSF (Pitrak et al. 1996). However, the physiological significance of the anti-apoptotic activity of G-CSF remains to be clarified.

PMN may also be capable of contributing to the elimination of G-CSF, as constant rate G-CSF infusions resulted in decreasing G-CSF serum levels when neutrophil counts increased (Morstyn et al. 1989).

2.2 Effects on Eosinophilic and Basophilic Granulocytes

To the author's knowledge, no literature is available on the effects of G-CSF on eosinophilic granulocytes. G-CSF had no effect on basophil migration or survival in vitro (Yamaguchi et al. 1992a and b).

3 Effects on Mononuclear Cells

The ability of G-CSF to enhance the functional activity of mononuclear cells was overlooked for a long time as its action on neutrophils was perceived as the most important immunomodulation. For this reason, the data concerning these cells are still sparse.

3.1 Effects on the Functions of Monocytes/Macrophages

A Synthesis of Mediators and Enzymes

G-CSF failed to induce IL-1ra protein production in monocytes stimulated with LPS or cultured on adherent IgG (Jenkins and Arend 1993), though neopterin production by the myelomonocytic cell line THP-1 in response to IFN-γ was increased in the presence of G-CSF (Marth et al. 1994). Furthermore, in contrast to the other three CSF discussed below, G-CSF did not induce any urokinase-type plasminogen activator (u-PA) activity (Hamilton et al. 1991). G-CSF had no effect on the synthesis of the components C3 or B in human monocytes (Høgåsen et al. 1993). Exposure of various rodent macrophage populations (Görgen et al. 1992) or human monocytes (Terashima et al. 1995) to G-CSF in vitro did not change TNF production induced by LPS, but its release was attenuated when neutrophils were additionally present in the assay (Terashima et al. 1995). In contrast, in our hands, highly purified human monocytes produced less LPS-inducible TNF-α in the presence of rhuG-CSF (unpublished observations).

When we measured the ex vivo LPS-stimulated TNF release of a number of different rodent macrophage populations prepared from donor animals pretreated in vivo with G-CSF, we found that TNF release was markedly suppressed compared with cells from control animals (Görgen et al. 1992). This was also the case when human whole blood of G-CSF-treated volunteers was employed in the assay (Hartung et al. 1995a; Baram et al. 1996) and held true also for a variety of stimuli other than LPS, such as preparations from Gram-positive bacteria, superantigens or phorbol ester (Hartung et al. 1995a). However, the ex vivo spontaneous TNF release by human monocytes from volunteers was the same, whether these were treated in vivo with G-CSF or not (Wiltschke et al. 1995). When volunteers were challenged in vivo with endotoxin 24 hrs after treatment with G-CSF, decreased TNF, IL-6 and IL-8 were found, while sTNF-R and IL-1ra were increased (Pajkrt et al. 1997).

These findings therefore confirm the antiinflammatory activity of G-CSF treatment observed ex vivo. However, when the endotoxin challenge was performed 2 h after injection of G-CSF, both pro- and anti-inflammatory factors were increased. Therefore, G-CSF required a certain time to establish an anti-inflammatory state of endotoxin responsiveness.

In some infection models, e.g. *E. coli* peritonitis in dogs (Eichacker et al. 1994) and cecal puncture in rats (Lundblad et al. 1996), G-CSF treatment also reduced the release of TNF in response to the infection. Unfortunately, it could not be determined whether less TNF was produced because the bacteria were eliminated faster by the increased numbers of granulocytes, or whether G-CSF had a direct effect on TNF production.

During G-CSF therapy of cancer patients after chemotherapy, serum TNF (3 out of 5 patients) and urinary peptido-leukotriene metabolites decreased (Denzlinger et al. 1994). In 18 patients with acute leukemia where G-CSF was administered in the recovery phase during 27 courses of consolidation chemotherapy, peripheral blood monocytes spontaneously produced high concentrations of IL-6 ex vivo (Liu et al. 1993). In 38 neutropenic gynecological cancer patients, G-CSF treatment doubled the concentration of serum neopterin (Marth et al. 1994), which is regarded as a marker of macrophage activation.

B
Expression of Surface Molecules

The expression of HLA-DR antigen was not affected by G-CSF in culture (Gerrard et al. 1990).

We measured a significant decrease of monocytic HLA-DR (–40%) on monocytes from G-CSF-treated volunteers (unpublished results). In contrast, monocytes in the blood of refractory testicular cancer patients treated with G-CSF, expressed more MHC-I, while MHC-II remained unaffected (Wiltschke et al. 1995). In our volunteer study, CD16 and CD64 on monocytes was significantly increased under G-CSF treatment.

C
Other

The release of reactive oxygen species was neither induced directly by G-CSF nor augmented when human monocytes were exposed to an exogenous signal such as fMLP or wheat germ agglutinin (Nathan 1989; Yuo et al. 1989). G-CSF had no effect on monocyte cytotoxicity toward WEHI fibrosarcoma cells (Cannistra et al. 1988) or toward the tumor cell line U937 (Wiltschke et al. 1995).

3.2 Effects on the Functions of Lymphocytes

To date, the effect of G-CSF on lymphocytes has found little attention. Since lymphocytes apparently have no G-CSF receptors, indirect effects via bystander cells, endogenous mediators or effects on lymphocyte precursors in bone marrow must be assumed.

A Mediator and Enzyme Synthesis

TNF secretion by human PBMC in response to allogeneic Daudi cells was suppressed by G-CSF in vitro (Kitabayashi et al. 1995). However, the production of IL-2 by lymphocytes in response to staphylococcal enterotoxin B (SEB) was not affected by the presence of G-CSF in vitro (Aoki et al. 1995). T-cells from G-CSF-treated normal mice showed reduced IL-2 and IFN-γ release while IL-4 formation was augmented (Pan et al. 1995). In contrast, in vivo G-CSF-treated galactosamine-sensitized mice were protected against T-cell mediated liver injury initiated by the superantigen SEB and exhibited reduced IL-2 serum levels without an effect on TNF release (Aoki et al. 1995).

When we treated human healthy volunteers acutely with G-CSF, we observed a reduced capacity of blood lymphocytes to release IFN-γ (Hartung et al. 1995a). On the one hand, this attenuated IFN-γ formation can be attributed to reduced TNF and IL-12 formation by monocytes from G-CSF-treated donors. On the other hand, this decrease in IFN-γ release was also noted in response to direct lymphocyte stimulators such as bacterial exotoxins and phorbol esters. In a recent study, we examined the effects of daily G-CSF treatment for a period of twelve days in 24 healthy volunteers. Compared to a placebo group, TNF-α, IL-12 and IFN-γ release of whole blood samples in response to ex vivo stimulation by LPS was reduced in each G-CSF group (75, 150 or 300 μg per day) throughout treatment. The presence of IL-12 added in vitro to LPS-stimulated blood from G-CSF-treated donors partly restored the attenuated IFN-γ and TNF-α release capacity, indicating that the suppression of IL-12 release is pivotal in the anti-inflammatory activity of G-CSF (submitted for publication). In this study, we also found that the release of IL-2 by ex vivo anti-CD3-stimulated whole blood or mononuclear cells was increased 24 h after the first injection. However, the increase in

IL-2 release was no longer significant after a further 24 h (submitted for publication). IL-2 may be involved in stimulating lymphocytosis which was observed in this treatment group.

Increased IL-2 formation and lymphocytosis might have therapeutic benefits especially for HIV-infected patients. These observations, therefore, prompted a further study, where we compared cytokine release of stimulated blood from volunteers and patients with advanced HIV infections and determined the effect of prior incubation with G-CSF in vitro. G-CSF did not have a suppressive effect on HIV-infected patient blood concerning LPS-induced TNF-α and IFN-γ formation, as it did in the blood of normal controls. However, G-CSF was able to partially restore impaired IL-2 but not IL-4 production by blood of HIV-infected patients in response to SEB (submitted for publication).

B
Expression of Surface Molecules

FcRII (CD32) can be induced by G-CSF on various human myeloid leukemia cell lines (KG-1, HL-60, U937, K562), though less potently than by GM-CSF (Liesveld et al. 1988). In 19 patients who had received high-dose chemotherapy (Dreger et al. 1993) and in 14 patients with myelodysplastic syndromes (Ganser et al. 1994), G-CSF or G-CSF combined with all-trans retinoic acid, respectively, resulted in an increase in the serum concentration of soluble IL-2 receptor, which is regarded a marker of T-cell activation. In a case of SCN, the level of endogenous G-CSF was elevated to 300 pg/ml before treatment and the lymphokines GM-CSF, IL-2 and IL-3 were slightly elevated. The patient showed only moderate response to high dose G-CSF [1600 μg/m^2/d], but a drastic increase in soluble IL-2 receptor was found in the absence of significant changes in the levels of lymphokines (Shitara et al. 1994).

C
Other

When we treated healthy volunteers for 12 consecutive days with G-CSF, a biphasic change in ex vivo blood lymphocyte proliferation inducible with either PHA or anti-CD3-antibodies was found: 24 h after the first injection, lymphocyte proliferation doubled but returned to pre-treatment values after 48 h; continuation of treatment led to suppression of lymphocyte proliferation. The clinical significance of this finding, e.g. in transplant rejection or autoimmune diseases is not yet clear.

In mice, delayed-type hypersensitivity to sheep erythrocytes was enhanced by G-CSF (Terashita et al. 1996).

4
Effects on the Functions of Other Cells

G-CSF acts mainly on leukocytes; the pertinent actions are summarized in Fig. 2. However, effects on other cells have also been noted as discussed in the following. G-CSF stimulated proliferation and migration of endothelial cells in vitro (Bussolino et al. 1989). G-CSF also augmented the secondary aggregation of platelets induced by a low concentration of adenosine diphosphate (Shimoda et al. 1993). The priming effect of G-CSF on platelets was specific to G-CSF as it could be prevented by anti-G-CSF antibody. Injection of G-CSF into mice induced histidine decarboxylase and ornithine decarboxylase in the spleen and bone marrow, suggesting that these two enzymes that catalyze the production of histidine and putrescine, respectively, may act as pacemakers in the early stages of hematopoiesis (Endo et al. 1992).

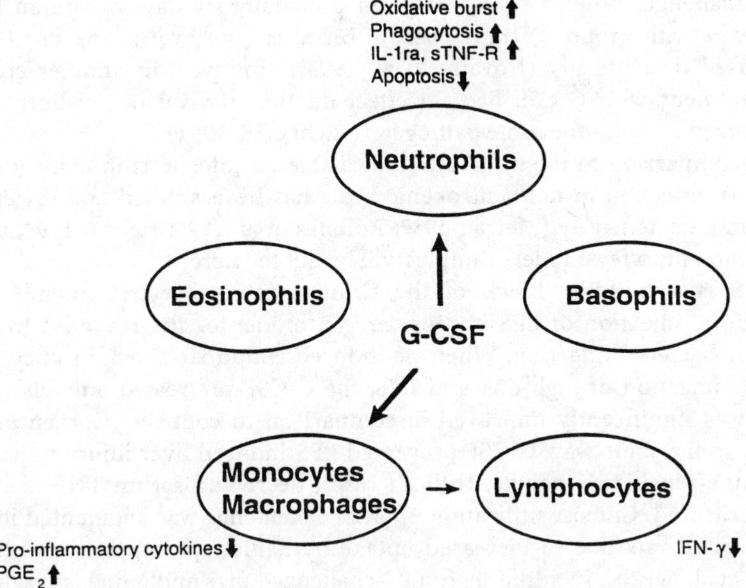

Fig. 2. Pertinent immunomodulatory effects of G-CSF on different leukocyte populations. The diagramm summarizes the predominant effects of G-CSF on immune functions of different leukocyte types

5 Infections

A Bacterial Infections

Improved neutrophil counts and functions through treatment with G-CSF were beneficial in a variety of infection models with absolute or functional neutropenia: Mice made neutropenic with cyclophosphamide and challenged with *Pseudomonas aeruginosa*, *Serratia marcescens* or *Staphylococcus aureus* (Matsumoto et al. 1987; Ono et al. 1988) had a higher survival rate when treated with G-CSF. G-CSF also improved recovery in a *Pseudomonas aeruginosa* burn wound infection model in mice (Mooney et al. 1988; Silver et al. 1989; Sartorelli et al. 1991), especially in combination therapy with gentamycin (Silver et al. 1989). G-CSF treatment was further beneficial to mice challenged with *Pseudomonas aeruginosa* pneumonia after hemorrhage (Abraham and Stevens 1992). Newborn rats known to have impaired PMN functions challenged with group B streptococci were protected (Cairo et al. 1990b and c; Cairo et al. 1992b) when G-CSF was given simultaneously with challenge. When G-CSF was given prenatally six days before an LD_{90} challenge with group B streptococci at birth, the survival of the pups was increased dramatically (Novales et al. 1993). However, in another study, pretreatment with G-CSF had no effect on the survival of newborn rats challenged i.p. with the same pathogen (Iguchi et al. 1991).

In comparison to the information available on infections in neutropenic animals, infection in nonneutropenic hosts has been studied and reviewed even more extensively (Metcalf 1987; Roilides et al. 1991; Nelson 1994; Dale 1994), so only a few model examples will be quoted here.

LPS is a building block of the Gram-negative bacterial membrane. Therefore, injection of LPS is considered a model for the response to unspecific bacterial infection. When we induced endotoxic shock in mice and rats by injection of high doses of LPS, the G-CSF-pretreated animals' survival was significantly improved in comparison to controls (Görgen et al. 1992). In the same way, G-CSF prevented LPS-induced liver injury in galactosamine-sensitized rodents. In both cases, decreased serum TNF activity was measured. Glucose utilization upon LPS challenge was augmented in G-CSF-treated rats due to increased uptake by ileum, spleen, liver and lung (Lang et al. 1992b). In addition, in LPS-challenged pigs and guinea pigs, lung injury was partially attenuated after G-CSF pretreatment (Kanazawa et al. 1988; Fink et al. 1993).

G-CSF also offered protection in a variety of peritonitis models: As little as 2 μg/kg administered after the onset of peritonitis brought about by cecal ligation and puncture was beneficial and synergized with antibiotics (O'Reilly et al. 1992; Goya et al. 1993), while a dose of 1 μg/kg G-CSF resulted in decreased lethality and improved histopathology in lung as well as renal and hepatic functions (Toda et al. 1993). Similarly, in a rat cecal puncture model, G-CSF decreased lethality and serum TNF levels (Lundblad et al. 1996). Rats were also protected by G-CSF in combination with antibiotics when a human stool suspension was inoculated into the peritoneum, in comparison to rats receiving only antibiotics (Lorenz et al. 1994). Here, a correlation between TNF levels and mortality rate was observed and G-CSF-treated animals were found to have total suppression of TNF serum levels. Benefits were obtained in analogous mouse and rat models without concurrent antibiotic therapy (Barsig et al. 1996; Dunne et al. 1996). Sepsis after intraperitoneal (i.p.) implantation of an *E. coli*-infected clot in dogs was improved by G-CSF (Eichacker et al. 1994).

In pneumonia models, G-CSF effects were not unequivocally advantageous: G-CSF pretreatment protected rats against pneumococcal pneumonia (Lister et al. 1993), but no effect was noted in rats fed a chronic ethanol diet, a model for immunosuppression experienced by alcohol abusers. In a similar experimental infection model (*Klebsiella pneumoniae*), however, G-CSF protected both control and ethanol-treated animals (Nelson et al. 1991). Rabbits inoculated transtracheally with *Pasteurella multocida* and administered G-CSF 24 h postchallenge tended towards improved survival (77 vs. 67%), but showed a slight increase in inflammation of liver, spleen and lung (Smith et al. 1995). Detrimental effects of G-CSF were described in rats treated before, during and after intrabroncheal application of *E. coli* with or without hyperoxia (Freeman et al. 1996). In a mouse model, G-CSF administered from 24 h before challenge to three days postchallenge was found to improve survival in splenectomized mice exposed to an aerosol of *Streptococcus pneumoniae*. G-CSF administered for two days after infection, was also greatly beneficial to the survival rate in a lethal model of soft tissue infection with *Pseudomonas aeruginosa* in mice (Yasuda et al. 1990).

Mice pretreated daily with a combination of clarithromycin and G-CSF for three days before intravenous (i.v.) challenge with *Mycobacterium avium* complex and treated further until sacrifice had significantly decreased levels of infection in spleen and lungs compared with those treated with either factor alone or control animals (Lazard et al. 1993).

SCID, i.e. mice devoid of lymphocytes, and normal mice inoculated i.v. with *Listeria monocytogenes*, then injected i.p. with G-CSF for five days, had lower counts of viable bacteria in the liver than non-treated mice. Possible contributions by activated neutrophils, but not γ/δ T-cells nor activated macrophages, to the augmentation of anti-listerial activity were shown in the SCID mice (Kayashima et al. 1993).

Taken together, G-CSF treatment was found to be beneficial in a broad variety of infection models. In general, it was necessary to initiate treatment concurrent with challenge or even as pre-treatment, which might be easily explained by the enormous subsequent endogenous production of G-CSF. On the other hand, only very few negative side-effects were recorded, which might in part be explained by the artificial nature of the infection models: When a large bacterial inoculum is injected into a body compartment, this does not reflect the pathophysiology of normal infections. After G-CSF pre-treatment, this bacterial inoculum will meet an already multiplied and alert army of leukocytes. Under these extreme circumstances, the fulminant stimulation and the amplified defense may result in damage to the host.

B
Viral, Fungal and Parasitic Infections

G-CSF was found to be effective for stimulation of nonspecific protection against Sendai virus infection in normal mice (Azuma et al. 1992). Mice made neutropenic with cyclophosphamide treatment were protected against challenge with *Candida albicans*, *Cryptococcus neoformans* or *Aspergillus fumigatus* by G-CSF (Matsumoto et al. 1987; Ono et al. 1988; Uchida et al. 1992; Hamood et al. 1994). Others however, noted no beneficial effect in neutropenic murine cryptococcosis (Polak-Wyss 1991a) and local candidosis (Polak-Wyss 1991b).

6
Potential Clinical Applications

To date, there is broad consensus on the clinical efficacy of G-CSF in neutropenic disorders such as iatrogenic neutropenia (due to chemotherapy, radiotherapy or myelosuppressive drugs), idiopathic neutropenia, leukemic neutropenia, refractory or aplastic anemia and agranulocytosis (Hollingshead and Goa 1991; Gabrilove 1992; Glaspy and Golde 1992; Roilides and Pizzo 1992 and 1993; Steward 1993; Dale 1994; Frampton et al. 1995). In phase III clinical studies of patients with chemotherapy-induced neutropenia, a decrease in the incidence of infection after standard regimens and fewer days with infectious and febrile neutropenic episodes during recovery from bone marrow transplantation occurred concomitantly with the amelioration of neutropenia. The decrease in morbidity was associated with shorter hospitalization times and reduced administration of parenteral antibacterial agents (Frampton et al. 1995). G-CSF may facilitate dose optimization and permit limited dose intensification of standard chemotherapy. Furthermore, G-CSF is effective in mobilizing peripheral blood progenitor cells for subsequent reinfusion. But, it is not yet firmly established, how great the effect of dose intensification through G-CSF or stem cell rescue is on remission rates or survival times (Frampton et al. 1995). In addition, G-CSF was able to partially restore defective neutrophil functions, including oxidative burst and bactericidal activity in PMN from children receiving chemotherapy for cancer to levels similar to normal PMN in vitro (Lejeune et al. 1996). There is also a potential value in treating patients with severe neutropenia and sepsis who are unresponsive to antibiotics with neutrophils collected from donors whose pools have been expanded by G-CSF (Grigg et al. 1996). Such infusions were successful in three patients without multi-organ dysfunction and prophylactically in two patients with localized fungal infections undergoing bone marrow transplantation (Grigg et al. 1996).

In contrast to in vitro findings, various phase III clinical studies on patients with AML discerned that G-CSF therapy had no effect on blast cell proliferation. There has also been no evidence of stimulation of non-myelogenous malignancies by G-CSF, although a number of neoplasms are known to express the G-CSF receptor. On the other hand, stimulation with hematopoietic growth factors could be exploited to increase cell-cycling of leukemic blasts, making them more susceptible to chemotherapy (Frampton et al. 1995). Although G-CSF is known to reduce the risk of infections and to improve the quality of life (Kalra et al. 1995; Welte and Dale 1996), the relationship between therapeutic G-CSF and the leukemic transformation in patients with SCN is unclear (Kalra et al. 1995). As no patients with cyclic or idiopathic neutropenia have been reported to have developed leukemia, it seems that G-CSF is not involved in the pathophysiology of leukemia (Welte and Dale 1996).

In patients with HIV infection, G-CSF has reversed or prevented neutropenia even during periods of full-dose myelotoxic therapy (Mitsuyasu 1994; Frumkin 1997). Furthermore, G-CSF has improved defective neutrophil function in vitro and in vivo in the setting of HIV infection (Frumkin 1997). A study with eight AIDS patients with serious infections and neutropenia treated with G-CSF achieved successful treatment of life-threatening bacterial infections by stimulating the neutrophil immune system (Hengge et al. 1992). As monocytes and macrophages are prone to infection by HIV, G-CSF is potentially more valuable in this situation than GM-CSF (Hengge et al. 1992). In a phase I/II trial with 22 AIDS patients, the combined therapy of G-CSF and erythropoietin improved neutropenia and anemia. Increases in neutrophil number, $CD4^+$ and $CD8^+$, lymphocyte proliferative response, bone marrow cellularity and hemoglobin levels were seen. Re-institution of zidovudine resulted in a decline in reticulocytes and hemoglobin in 8 of 20 patients. Limiting dilution plasma and lymphocyte co-cultures for HIV as well as serum p24 antigen levels did not change significantly during G-CSF or combined G-CSF and erythropoietin therapy; HIV p24 antigen decreased significantly with zidovudine therapy. Opportunistic infections, which still occurred in 14 patients, were all successfully treated with myelosuppressive antimicrobial agents without the development of neutropenia (Miles et al. 1991). In animal models, G-CSF has been beneficial against opportunistic pathogens common to the HIV-infected population, though the clinical relevance of these findings remains to be explored (Frumkin 1997).

Diabetic patients are prone to infection characterized by a lack of neutrophilia and neutrophil dysfunction. The typical foot ulcers of these patients were efficiently treated compared to placebo treated controls by systemic application of G-CSF for 7 days (Gough et al. 1997). Concomitantly, superoxide formation by neutrophils of these patients was improved.

A promising indication for the use of G-CSF in non-neutropenics is sepsis prophylaxis and treatment in cases where the time point of a risk is known (major surgery) or can be anticipated (trauma, burn, local infection). For instance, 756 patients enrolled in a double-blind controlled multi-center study with community acquired pneumonia showed reduced incidence of adult respiratory distress syndrome (ARDS) and disseminated intravascular coagulation (DIC) (Andresen and Movahhed 1998). In another investigational study, 37 liver allograft recipients treated with G-CSF showed decreased septic episodes and less sepsis-associated death compared with 49 patients who did not receive G-CSF treatment (Foster et al. 1995). In 42 human neonates with presumed bacterial sepsis treated in a randomized controlled study with G-CSF, increased PMN counts had no association with any acute toxicity (Gillan et al. 1993 and 1994). It was proposed that neutropenia and deficient G-CSF production may predispose newborn infants to sepsis (25–30% incidence in neonates weighing 500 to 1000 g). However, there are still doubts whether respiratory distress syndrome in neonates is an indication for G-CSF (Yurdakök et al. 1994).

In another study ten postoperative/posttraumatic patients, who are known to be prone to sepsis due to an immunocompromized anergic state, were treated with G-CSF [1 μg/kg/d] for four days: none of these patients developed sepsis (Weiss et al. 1994 and 1995, Gross-Weege et al. 1997). The efficacy of G-CSF therapy was also investigated in 24 patients with neutropenia and sepsis who had failed to respond to antibiotics. In this study, 19 patients responded to daily injection of 75 μg G-CSF s.c. with an increase in

PMN and all survived. The five patients with no response to G-CSF died (Endo et al. 1994).

G-CSF may also have a therapeutic bearing for inflammatory diseases. Pretreatment of rabbits before or at induction of experimental immune complex colitis with G-CSF had anti-inflammatory effects, such as increased tissue MPO levels, decreased LTB_4 and thromboxane B_2 and unaffected PGE_2 dialysis fluid levels in comparison to untreated controls (Hommes et al. 1996).

In addition, G-CSF was able to lower the incidence of chronic graft-versus-host-disease (GVHD) in mice and patients receiving allogeneic bone marrow transplants (Hirokawa et al. 1995). It was observed that patients receiving G-CSF had a lower incidence of increased TNF-α serum levels than their counterparts not given G-CSF (Hirokawa et al. 1995). On the other hand, recipient survival was significantly improved in a murine acute GVHD model when the donors were pretreated with G-CSF (Pan et al. 1995). The advantage is thought to lie in the reduction of the secretion of IL-2 and IFN-γ and stimulation of the production of IL-4 by T-lymphocytes, a pattern which was observed in vitro. Thirteen days after transplantation, the splenocytes of recipients of G-CSF-treated bone marrow showed an increase in IL-4 along with a decrease in IL-2 and IFN-γ production.

There are indications that patients with glycogen storage disease (type 1b), who have experienced severe and/or recurrent infections, may benefit from G-CSF therapy (Frampton et al. 1995). G-CSF may also find a role in immunotherapy directed toward the proto-oncogene product HER-2/neu which is overexpressed in some breast cancers and other carcinomas: whole blood or isolated neutrophils from patients or volunteers were highly cytotoxic toward HER-2/neu breast cell carcinoma lines in the presence of a bispecific [HER-2/neu x CD64] antibody. G-CSF application not only increased effector cell numbers, but also induced their expression of CD64 (Repp et al. 1995; Stockmeyer et al. 1997). Similar observations were made with bispecific [HLA class II x CD64] antibodies toward malignant B-cell lines with G-CSF-primed human neutrophils in vitro (Elsasser et al. 1996).

The results of these orientating clinical investigations support the notion that an overwhelming concordance exists between the hitherto available set of preclinical data and the practical experiences gained by the world-wide usage of the drug. It therefore seems appropriate to explore the use of the pharmacological potential of G-CSF further for new indications such as non-neutropenic infection, sepsis prophylaxis, chronic inflammation and AIDS.

II
GM-CSF

1
General Information

1.1
Molecular Biology and Endogenous Production

A
The Molecule GM-CSF

GM-CSF was purified from mouse lung-conditioned medium in 1977 and from the supernatant of a HTLV I infected T-cell line (Mo) in 1984 (Freund and Kleine 1992). Murine GM-CSF was cloned in 1984 (Gough et al. 1984) and human GM-CSF in 1985 (Wong et al. 1985; Cantrell et al. 1985). Gasson et al. first demonstrated that NIF-T (neutrophil migration inhibition factor) and GM-CSF were one and the same protein, which acts on progenitors as well as mature cells of the granulocyte and monocyte lineage (Gasson et al. 1984).

The single copy of the gene encoding human GM-CSF is localized on chromosome 5, region 5q21–5q32 (Huebner et al. 1985), close to the genes for IL-3, IL-4, IL-5, M-CSF, the M-CSF receptor and an early growth response factor (EGR-1). Interstitial deletions in this region are seen predominantly in therapy-related myelodysplastic syndromes and acute leukemias as well as in patients with deletions in the 5q region, who display a refractory anemia with morphologic abnormalities of the megakaryocytes (Gasson 1991). It is speculated that the deletion of this region unmasks aberrant or recessive alleles that are involved in the generation of abnormal hematopoiesis, or that the deletion of an important suppressor gene, which may be closely linked to these hematopoietic regulators, is the cause.

The gene for GM-CSF encodes a protein of 144 amino acids (Cantrell et al. 1985), which undergoes cleavage of a 17-amino acid segment from the amino terminus, resulting in a mature protein of 127 amino acids (Ruef and Coleman 1990), with four α-helices and 2 β-sheets, stabilized in a bilobed configuration by two disulfide bridges. The apparent range of molecular mass lies between 14’500 and 35’000 (Clark 1988). This heterogeneity is due to the variable N-linked addition of complex carbohydrate at two sites and the glycosylation of two serine residues near the amino terminus.

Although non-glycosylated GM-CSF expressed in *E. coli* (Molgramostim) and glycosylated GM-CSF expressed in Chinese hamster ovary (CHO) carcinoma have comparable effects on the induction of leukocytosis in monkeys in vivo (Mayer et al. 1987), their pharmakokinetics, as studied in healthy volunteers, were quite different: the serum concentration of the non-glycosylated form reached a higher maximum faster and also decreased more rapidly than the glycosylated form (Denzlinger et al. 1993). The two forms may be distributed differently in the body, in which case they would be differentially effective in reaching the various populations of target cells.

Preclinical studies with recombinant human GM-CSF are limited by the lack of cross species reactivity in mice (Metcalf 1986). The protein sequence homology between human and murine GM-CSF is only 60% (Wong et al. 1985). Human GM-CSF does not appear to activate canine PMN in vitro and may actually down-regulate their inflammatory responses (D'Alesandro et al. 1991). Even in monkeys, only short-term studies can be undertaken because antibodies develop toward human GM-CSF (Morstyn et al. 1989).

B
Endogenous Production of GM-CSF

The expression of GM-CSF by T-lymphocytes (Herrmann et al. 1988; Bickel et al. 1990), B lymphocytes (Akahane et al. 1991), fibroblasts (Nimer et al. 1989), mast cells (Wodnar-Filopowicz et al. 1989; Gasson 1991), endothelial cells (Sieff et al. 1987a; Gasson 1991), epithelial cells (Galy and Spits 1991; Churchill et al. 1992; Ohtoshi et al. 1994), macrophages (Akahane and Pluznik 1993), NK cells (Cuturi et al. 1989; Levitt et al. 1991) and even cultured astrocytes (Ohno et al. 1990) is controlled by either transcriptional or posttranscriptional regulation: neither T-cells, macrophages nor NK cells produced stable cytoplasmic mRNA in the absence of stimulating signals in culture conditions (Oster et al. 1989c; Levitt et al. 1991). In the case of the T-cells and mesothelial cells, the upregulation occured through the increase of GM-CSF mRNA production (Di Persio and Abboud 1992). In contrast, the activation of fibroblasts, endothelial cells and macrophages resulted in increased stability of the mRNA and therefore increased protein synthesis (Munker et al. 1986; Koeffler et al. 1987; Nimer et al. 1989; Di Persio and Abboud 1992). Obviously, both transcriptional and posttranscriptional signals act together to modulate the level of GM-CSF mRNA in NK cells (Levitt et al. 1991). The secretion of GM-CSF protein was increased greatly on maturation of monocytes to macrophages in vitro (Krause et al. 1992; Kruger et al. 1996a). Two of four tested malignant mesothelioma cell lines expressed GM-CSF transcripts autonomously (Demetri et al. 1989). A case of a thyroid cancer producing GM-CSF and G-CSF autonomously has been reported (Nakada et al. 1996).

Induction of GM-CSF production

Table 6 provides an overview of the factors involved in initiating the production of GM-CSF. Monocytes/macrophages were directly activated by immune or inflammatory stimuli, such as LPS (Thorens et al. 1987; Lee et al. 1990; Sallerfors and Olofsson 1992; Hamilton et al. 1992b; Kruger et al. 1996a), IL-1 (Sallerfors and Olofsson 1992), PHA (Imakawa et al. 1993), fetal calf serum, thioglycolate broth as well as phagocytosis or adherence to fibronectin (Thorens et al. 1987) to produce GM-CSF protein. IL-3 induced a minor secretion of protein, but TNF, G-CSF and M-CSF were unable to induce the secretion of GM-CSF (Sallerfors and Olofsson 1992). There are conflicting reports as to the actions of IFN-γ and M-CSF. One study states that neither had any effect on GM-CSF production in monocytes (Sallerfors and Olofsson 1992), while one found that GM-CSF protein was produced by human monocytes stimulated by M-CSF (Motoyoshi et al. 1989). Another investigation found that concentrations of IFN-γ as low as 10 U/ml were able to induce the secretion of GM-CSF in these cells (Herrmann et al. 1986). However, the same study found that the response to IFN-γ was biphasic, i.e. concentrations greater than 250 U/ml did not induce detectable GM-CSF activity, as a humoral G/M-progenitor cell inhibitor was released that masked the effect of GM-CSF (Herrmann et al. 1986). The smooth-domed opaque variants of *Mycobacterium avium* induced the secretion of GM-CSF in infected macrophages, which seemed to suppress their own proliferation. In comparison, the smooth transparent variant of this strain, which proliferated better, did not induce the production of GM-CSF (Fattorini et al. 1994).

PBMC produced GM-CSF in response to PHA plus PMA (Oster et al. 1989c). T-lymphocytes produced GM-CSF when stimulated by IL-2 (Oster et al. 1989c), ConA ± PMA (Borger et al. 1996) or staphylococcal enterotoxin A (SEA), but not LPS or IL-1 (Sallerfors and Olofsson 1992). However, another study found that T-lymphocytes do produce GM-CSF when stimulated with IL-1 (Herrmann et al. 1988). Phorbol ester has been shown to induce GM-CSF production by murine T-lymphocytes (Bickel et al. 1990). T-cells were found to be able to differentially secrete GM-CSF and IL-3 independently (Fitzpatrick and Kelso 1995). GM-CSF production by NK cells was regulated by the IL-2Rβ and CD2 receptor, but not by IL-2Rα (Levitt et al. 1991). Both rIL-2 and CD16 ligands induced accumulation of GM-CSF mRNA in NK cells and the combined effect of the two stimuli was synergistic (Cuturi et al. 1989). As these cells are part of the first line of defense against infection and

Table 6. Factors affecting the production of the GM-CSF protein

Stimulus type	Stimulus	Cell type	Result	References
Bacterial molecules	LPS	mono/macro	+	Thorens et al. 1987
		B cells	+	Akahane et al. 1991
		T cells	–	Sallerfors and Oloffson 1992
		astrocytes	+	Ohno et al. 1990
	SEA	T cells	+	Sallerfors and Oloffson 1992
	Mycobacterium avium	mono/macro	+	Fattorini et al. 1994
	invasive bacteria	epithelium	+	Jung et al. 1995
	non-invasive bacteria	epithelium	–	Jung et al. 1995
	G. lamblia	epithelium	–	Jung et al. 1995
Pro-inflammatory mediators	IL-1	mono/macro	+	Sallerfors and Olofsson 1992
		T cells	+	Herrmann et al. 1988
			–	Sallerfors and Oloffson 1992
		B cells	+	Akahane et al. 1991
		endothelium	+	Sieff et al. 1987b
		mesothelium	+	Lanfrancone et al. 1992
		fibroblasts	+	Nimer et al. 1989
		smooth muscle	+	Filonzi et al. 1993
		epithelium	+	Galy and Spits 1991
	TNF-α	mono/macro	–	Sallerfors and Oloffson 1992
		endothelium	+	Broudy et al. 1986
		mesothelium	+	Demetri et al. 1989
		fibroblasts	+	Nimer et al. 1989
		smooth muscle	–	Filonzi et al. 1993
		epithelium	+	Jung et al. 1995
	IFN-γ	mono/macro	+	Herrmann et al. 1986
			–	Sallerfors and Oloffson 1992
		endothelium	–	Ross and Koeffler 1992
	IL-2	T cells	+	Oster et al. 1989c
		endothelium	–	Ross and Koeffler 1992
	IgE	mast cells	+	Wodnar-Filipowicz et al. 1989

Table 6 (continued)

Stimulus type	Stimulus	Cell type	Result	References
Colony stimulating factors	IL-3	mono/macro	+	Sallerfors and Oloffson 1992
	G-CSF	mono/macro	–	Sallerfors and Oloffson 1992
	M-CSF	mono/macro	+	Motoyoshi et al. 1989
			–	Sallerfors and Oloffson 1992
Other	PHA	mono/macro	+	Imakawa et al. 1993
	PHA + PMA	PBMC	+	Oster et al. 1989c
	Con A	T cells	+	Borger et al. 1996
	Con A + PMA	T cells	+	Borger et al. 1996
	Phorbol ester	T cells	+	Bickel et al. 1990
	Phorbol diester	fibroblasts	+	Nimer et al. 1989
	fetal calf serum	mono/macro	+	Thorens et al. 1987
	thioglycolate broth	mono/macro	+	Thorens et al. 1987
	phagocytosis	mono/macro	+	Thorens et al. 1987
	adherence to fibronectin	mono/macro	+	Thorens et al. 1987
	airpollutants	epithelium	+	Ohtoshi et al. 1994
	IL-4	epithelium	+	Nakamura et al. 1996
	IL-13	epithelium	+	Nakamura et al. 1996
	IL-10	epithelium	–	Nakamura et al. 1996
	ionomycin	eosinophils	+	Kita et al. 1991
		PMN	+	Cassatella 1995

Key: mono/macro : monocytes or macrophages
+ : induction of GM-CSF production and secretion
– : no effect on the production of GM-CSF

tumor metastasis, their secretion of this factor may contribute to the activation of macrophages and neutrophils and to the recruitment of other effector cells (Ross and Koeffler 1992).

Other leukocytes may also be stimulated to produce GM-CSF: Murine B lymphocyte cell lines were induced to express GM-CSF by IL-1α or LPS (Akahane et al. 1991). CD23 (IgE)-mediated activation of murine mast cells in vitro, as it happens in allergic diseases in vivo, resulted in the production of GM-CSF (Wodnar-Filopowicz et al. 1989), which may be interpreted as a mechanism of local tissue defense. Human peripheral blood eosinophils can be stimulated to release GM-CSF by ionomycin in vitro, suggesting a novel role for eosinophils in the pathophysiology of allergic inflammation and host-defense mechanisms (Kita et al. 1991). When patients with asthma were challenged with allergen inhalation, peripheral blood eosinophils expressed no GM-CSF mRNA or protein. Instead, both bronchoalveolar lavage eosinophils and mononuclear cells expressed GM-CSF mRNA and protein (Sullivan and Broide 1996).

In addition, PMN produced GM-CSF when stimulated by ionomycin in vitro. Even though they synthesize far smaller quantities of the cytokine than PBMC stimulated by PMA under similar conditions on a single cell basis, one must consider that granulocytes constitute the majority of infiltrating cells in inflamed tissues and may thus represent an important source of cytokines in such tissues (Cassatella 1995).

Apart from the cells of the immune system, various other tissues also have the ability to produce GM-CSF in response to an exogenous signal or the mediators of inflammation. TNF-α and IL-1 stimulated the production of GM-CSF by human endothelial cells in vitro (Broudy et al. 1986; Munker et al. 1986; Sieff et al. 1987a; Seelentag et al. 1987; Zsebo et al. 1988; Akahane and Pluznick 1993), apparently via different mechanisms, as they were found to have an additive effect (Seelentag et al. 1987). On the contrary, IFN-γ and IL-2 had no effect on GM-CSF production by human umbilical cord endothelial cells (Ross and Koeffler 1992). GM-CSF mRNA transcripts in normal human mesothelial cells in vitro could be induced by a combination of EGF and TNF (Demetri et al. 1989) or by IL-1 (Lanfrancone et al. 1992). Internal mammary artery and aortic smooth muscle cells produced GM-CSF when stimulated by IL-1 or TNF-α (Filonzi et al. 1993). Fibroblasts increased their production of GM-CSF upon exposure to TNF (Munker et al. 1986; Koeffler et al. 1987; Nimer et al. 1989) or IL-1 (Nimer et al. 1989). These two cytokines may synergize in this respect (Seelentag et al. 1989). Phorbol diester also stimulated fibroblasts to produce GM-CSF (Nimer et al. 1989). Monolayers of human colon epithelial cell lines (T84, HT29, Caco-2) upregulated GM-CSF expression and release when infected with invasive strains of bacteria

(*Salmonella dublin*, *Shigella dysenteriae*, *Yersinia enterocolitica*, *Listeria monocytogenes*, enteroinvasive *E. coli*) or when stimulated by TNF-α or IL-1 (Jung et al. 1995). In contrast, cytokine gene expression was not altered after infection of these cells with noninvasive bacteria or the noninvasive protozoan parasite *Giardia lamblia* (Jung et al. 1995). Human thymic epithelia also upregulated their production of GM-CSF in response to IL-1 (Galy and Spits 1991). Human airway epithelial cells in culture released GM-CSF in response to a wide variety of air pollutants (Ohtoshi et al. 1994); the subsequent actions of GM-CSF on eosinophils was suggested to explain the recent increase in the prevalence of allergic disorders. GM-CSF production by bronchial epithelial cells of patients with respiratory diseases may also be upregulated by IL-4 or IL-13, but not by IL-10 (Nakamura et al. 1996). Furthermore, GM-CSF mRNA has been found localized in the luminal and glandular epithelium of the uterine endometrium (Imakawa et al. 1993). Cultured astrocytes released GM-CSF in response to stimulation by LPS (Ohno et al. 1990).

GM-CSF production has been reported from different cell sources in some types of disease: For example, leukemic cells from some patients with AML secreted cytokines including TNF and IL-1, which in turn stimulated accessory cells to produce, among others, GM-CSF (Oster et al. 1989a; Ross and Koeffler 1992). Several AML patients have been identified, whose blast cells are autocrine for GM-CSF, suggesting that in a population of leukemic cells, there may be a selective advantage in producing GM-CSF (Clark and Kamen 1987). When T-cells are activated in some autoimmune diseases, the observed eosinophilia may in part be due to the production of GM-CSF (Gabrilove and Jakubowski 1990).

In conclusion, GM-CSF is produced by a broad variety of cells exceeding the immune cells. Since most inflammatory stimuli and mediators are capable of inducing GM-CSF formation, the presence of GM-CSF at any inflammatory focus must be assumed. Notably, systemic GM-CSF levels are found only rarely, which indicates a primarily auto- and paracrine role for this local GM-CSF formation.

Modulation of GM-CSF production

In the presence of LPS, GM-CSF production by purified human monocytes was upregulated by cyclooxygenase inhibition (indomethacin); this action was reversed by exogenous PGE_2 (Lee et al. 1990; Hamilton 1994). The accumulation of GM-CSF mRNA in macrophages was prevented by the corticosteroids, e.g. dexamethasone and IFN-γ (Thorens et al. 1987; Hamilton et al. 1992b; Lenhoff and Olofsson 1996). Exogenous as well as endogenous IL-10 has also been reported to inhibit the expression of the GM-CSF gene in stimulated monocytes (Kruger et al. 1996b). Human blood monocytes infected in vitro with HIV-1 downregulated their GM-CSF production (Esser et al. 1996). IL-4, a T-cell-produced cytokine, had an inhibitory function on GM-CSF production in LPS-stimulated monocytes (Hamilton et al. 1992b) and in normal PBMC (Sallerfors 1994). In an autologous mixed lymphocyte reaction in human PBMC, IL-10 suppressed the production of GM-CSF (Sagawa et al. 1996). IFN-α inhibited the gene expression of GM-CSF in PBMC, whether mitogen- or antigen-induced (Krishnaswamy et al. 1996). 1,25 dihydroxyvitamin D_3 (Tobler et al. 1987) as well as corticosteroids, cyclosporin and cephalosporins (Lenhoff and Olofsson 1996), inhibited the GM-CSF production by stimulated T-lymphocytes in vitro. Furthermore, activation of the cAMP-dependent signaling pathway, e.g. by 2'-O-dibutyryl-cAMP (db-cAMP), PGE_2 or isoproterenol, reduced the secretion of GM-CSF protein in ConA ± PMA-activated T-cells (Borger et al. 1996). TNF-α or IFN-γ, which synergize at suboptimal concentrations, enhanced the release of GM-CSF by PHA-stimulated subsets of T-lymphocytes in vitro (Lu et al. 1988b). IL-4 also inhibited GM-CSF production by IL-1α-stimulated murine B cells (Akahane and Pluznik 1992).

The secretion of GM-CSF by endothelial cells was downregulated by cyclosporin and cephalosporins (Lenhoff and Olofsson 1996). IFN-γ downregulated the production of GM-CSF by murine vascular endothelial cells by destablizing the mRNA (Akahane and Pluznick 1993) and by TNF- or IL-1-stimulated fibroblasts (Hamilton et al. 1992a). Similarly, dexamethasone (Hamilton et al. 1992a; Sallerfors 1994) or 1,25 dihydroxyvitamin D3 (Sallerfors 1994) or IL-4 (Hamilton et al. 1992a) decreased, while basic fibroblast growth factor and cyclooxygenase inhibitors, although nonstimulatory themselves, potentiated GM-CSF release by fibroblasts (Hamilton et al. 1992a). Both IFN-γ and IL-4 inhibited IL-1-stimulated GM-CSF production in thymic epithelial cells (Galy and Spits 1991), and dexamethasone as well as steroids and anti-allergic drugs, lowered GM-CSF production (Churchill et al. 1992; Ohtoshi et al. 1994), but histamine as well as IL-1 enhanced the release of GM-CSF by these cells (Churchill et al. 1992).

An amplification network, i.e. positive feedback loops, seems to exist at many sites the immune response, as IL-1 and GM-CSF are produced by many of the same cells and GM-CSF can induce secretion of both IL-1 and TNF (Ruef and Coleman 1990). Conversely, TNF produced for instance by macrophages, causes the secretion of GM-CSF by a variety of cells (Munker et al. 1986). The reader is referred to Table 7 for a summary of these interactions. In general, anti-inflammatory cytokines, such as IL-4 and IL-10, appear to attenuate GM-CSF formation, but the pro-inflammatory IFN-γ has similar effects in some cells.

Table 7. Modulation of GM-CSF production in stimulated cells

Cells	Augmenting factors	References	Inhibiting factors	References
Monocytes/ Macrophages	indomethacin	Lee et al. 1990	dexamethasone IFN-γ IL-4 IL-10 PGE, HIV infection	Thorens et al. 1987 Thorens et al. 1987 Hamilton et al. 1992b Kruger et al. 1996b Lee et al.1990 Esser et al. 1996
PBMC			IFN-α IL-4 IL-10	Krishnaswamy et al. 1996 Hamilton et al. 1992b Sagawa et al. 1996
T cells	TNF-α IFN-γ	Lu et al. 1988b Lu et al. 1988b	1,25(OH), Vitamin D3 corticosteroids cyclosporins cephalosporins db-cAMP PGE, isoproterenol	Tobler et al. 1987 Lenhoff and Olofsson 1996 Lenhoff and Olofsson 1996 Lenhoff and Olofsson 1996 Borger et al. 1996 Borger et al. 1996 Borger et al. 1996
B cells			IL-4	Akahane and Pluznik 1992
endothelium			cyclosporins cephalosporins IFN-γ	Lenhoff and Olofsson 1996 Lenhoff and Olofsson 1996 Akahane and Pluznick 1993
fibroblasts	basic fibroblast growth factor cyclooxygenase inhibitor	Hamilton et al. 1992a Hamilton et al. 1992a	IFN-γ dexamethasone 1,25(OH), Vitamin D3 IL-4	Hamilton et al. 1992a Hamilton et al. 1992a Sallerfors 1994 Hamilton et al. 1992a
epithelium	histamine IL-1	Churchill et al. 1992 Churchill et al. 1992	IFN-γ IL-4 dexamethasone steroids anti-allergic drugs	Galy and Spits 1991 Galy and Spits 1991 Churchill et al. 1992 Ohtoshi et al. 1994 Ohtoshi et al. 1994

Serum GM-CSF

GM-CSF has rarely been found in the circulation at detectable levels, but is extractable from all major organs at higher concentrations (Metcalf 1987). It usually acts locally in a paracrine manner (Gasson 1991).

Mice infected with *Listeria monocytogenes* produced only small amounts of GM-CSF measurable in the serum (Cheers et al. 1988). The blood of patients with experimental endotoxemia (Granowitz et al. 1992), neutropenic fever (Cebon et al. 1994), cyclic neutropenia (Misago et al. 1991), sepsis (Cebon et al. 1994) or acute lymphoblastic leukemia (ALL) or acute undifferentiated leukemia (AUL) (Sallerfors and Olofsson 1991) also did not show elevated GM-CSF serum levels.

The cerebrospinal fluid of patients with meningitis may contain a measurable concentration of GM-CSF, emphasizing a role for local production of the CSF (Sallerfors 1994). An increase in plasma GM-CSF in conjunction with infections in immunosuppressed renal transplant patients as well as in patients with inflammatory disorders such as asthma (Sallerfors 1994), AML, aplastic anemia and granulocytosis due to infection (Omori et al. 1992) has been described. A woman with a thyroid carcinoma, which produced GM-CSF and G-CSF autonomously, had significantly elevated serum levels of both factors without any evidence of infection (Nakada et al. 1996).

Deficient GM-CSF

Mice with homozygous mutations of the GM-CSF gene showed no major deficits in hematopoiesis until 12 weeks of age, but they developed abnormal lungs with extensive peribronchovascular infiltration of lymphocytes, especially B cells; the alveoli contained granular eosinophilic material and lamellar bodies indicative of surfactant accumulation, and there were numerous large intra-alveolar phagocytic macrophages. Some mice had sub-clinical bacterial or fungal infections (Stanley et al. 1994). These observations indicate that GM-CSF is not essential for the maintenance of hematopoietic cells and their precursors, but rather for normal pulmonary physiology and resistance to local infection (Stanley et al. 1994).

Mice deficient in both GM-CSF and M-CSF displayed the sum of the features observed in mice lacking production of either protein alone: they had osteopetrosis, were toothless and developed a more severe form of the lung pathology described above. Their survival was significantly reduced compared to that of mice deficient in only one of the factors and all mice had pneumonia at death. The diseased lungs contained numerous phagocytically active macrophages and the levels of circulating monocytes were comparable to those of M-CSF-deficient mice (Lieschke et al. 1994b).

Excess GM-CSF

Administration of a high dose of canine GM-CSF to dogs for twelve days reduced the survival of labeled platelets measured in vivo, by increasing the hepatic uptake and destruction of platelets. Platelets were identified in association with Kupffer cells in the liver, which might be due to increased expression of CD1c and CD11c on these liver macrophages (Nash et al. 1995).

GM-CSF caused increased adhesion of monocytes to endothelial cells (Gamble et al. 1989), which implies that egress of monocytes into the tissue spaces is also enhanced. This may also be the case with neutrophils via TNF production stimulated in monocytes by GM-CSF (Cannistra et al. 1987). These effects combined with the subtle activation of PMN by GM-CSF may combine to produce excessive tissue infiltration and inflammation. Inappropriate or excessive production of GM-CSF was examined in transgenic mice which carried the murine GM-CSF gene expressed via a retroviral promoter and exhibited high serum levels of GM-CSF. Accumulations of macrophages in the eyes, striated muscle and peritoneal and pleural cavities were observed, leading to blindness and muscle wasting (Lang et al. 1987). The engraftment of bone marrow cells infected with a recombinant retrovirus, MPZen (GM-CSF), into irradiated mice also resulted in very high levels of serum and tissue GM-CSF. The mice died within 4 weeks of transplantation with extensive neutrophil and macrophage infiltration of the spleen, lung, liver, peritoneal cavity, heart and skeletal muscles. Furthermore, the thymus and lymph nodes were deficient in lymphoid cells. Yet, no disease occurred when infected cells from hematopoietic tissues of the primary transplanted animals were injected into normal or sub-lethally irradiated mice (Johnson et al. 1989). Highly expressed GM-CSF in rat lung, after intrapulmonary transfer of the gene coding for murine GM-CSF using an adenoviral vector, led to the sustained but self-limiting accumulation of eosinophils and macrophages associated with tissue injury in the lung. After this, varying degrees of irreversible fibrotic reactions were observed in later stages, suggesting that GM-CSF plays a role in the development of respiratory conditions characterized by eosinophilia, granuloma and/or fibrosis (Xing et al. 1996). There is no evidence from murine studies that the overproduction of GM-CSF in transgenic mice has any effect on leukemic transformation, even when an attempt was made to induce leukemia by irradiation (Scarffe 1991).

The in vivo administration of GM-CSF to a severely pancytopenic monkey infected with simian type D retrovirus caused not only leukocytosis, but also a substantial reticulocytosis, which might be explained by a synergy between GM-CSF and the high circulating levels of erythropoietin measured in this monkey in comparison to a control animal (Donahue et al. 1986). High doses of human GM-CSF given to the monkey resulted in generalized polyserositis, hepatic necrosis and pancellular hyperplasia of the bone marrow.

In conclusion, a variety of pronounced side-effects were associated with high GM-CSF levels in a number of animal studies. However, at lower, pharmacologically active doses, GM-CSF was generally well tolerated in both rhesus monkeys and humans (Robinson and Myers 1993).

Toxicity

Adverse effects of GM-CSF are more common and more severe than those of G-CSF. Toxicity is dose related and greater with i.v. than s.c. administration (Frank and Mandell 1995): First-dose reactions may consist of flushing, tachycardia, hypotension, hypoxia, myalgias and vomiting. At high doses in humans, toxicities such as bone pain, rash, fever and chills, myalgias, phlebitis, thrombosis, capillary leak syndrome, edema, lethargy, anorexia and pericardial and pleural effusions were reported (Di Persio 1990; Frank and Mandell 1995). It is difficult to determine the frequency of the side-effects attributable to GM-CSF, because many of these symptoms are common after bone marrow transplantation: There are also case reports in which GM-CSF appeared to reactivate autoimmune thyroiditis, rheumatoid arthritis and hemolysis. Thus, GM-CSF should be used cautiously in patients with a history of autoimmune or chronic inflammatory disease (Lieschke and Burgess 1992b).

C
Receptors and Signal Transduction

GM-CSF receptors are expressed on human neutrophils, eosinophils and monocytes, on purified CFU-GM and on some non-hematopoietic cells (Di Persio et al. 1988a). Neutrophils displayed only a single class of high-affinity receptors, while monocytes and myeloid leukemic cell lines had two classes of binding sites, one with high and one with lower affinity (Moore 1991). The cDNA for two receptor subunits, designated α and β, have been cloned (Gearing et al. 1989; Hayashida et al. 1990). The α-subunit cDNA encodes a low-affinity binding protein for GM-CSF, but cotransfection of this subunit with the β-subunit cDNA reconstituted the high-affinity GM-CSF-R (Hayashida et al. 1990). Isolated expression of the β-subunit, which is shared with the receptors for IL-3 and IL-5 (Kitamura et al. 1991; Tavernier et al. 1991), did not generate any binding capacity for human GM-CSF.

The high-affinity receptor has a molecular weight of 84'000 (Di Persio et al. 1988a) with an affinity of approximately 30 pM (Weisbart et al. 1989). It is expressed in low numbers (50–1000 sites per cell) (Rapoport 1992). The α-subunit was down-regulated by GM-CSF, PMA and calcium ionophore A23187 (Rapoport 1992).

GM-CSF receptors are expressed as soluble forms as well as membrane-anchored proteins (Lopez et al. 1992); the soluble isoforms arise from translation of specific mRNA and are the product of differential splicing.

GM-CSF had no direct effect on membrane depolarization, Ca^{2+} flux or intracellular free Ca^{2+} concentration (Sullivan et al.1987; Rapoport 1992) and produced no detectable increase in neutrophilic inositoltriphosphate or diacylglycerol levels, making phospholipase C activation an unlikely second messenger system (Rapoport 1992). Though there is no region on the low-affinity human GM-CSF receptor that would be expected to have tyrosine kinase activity (Rapoport 1992), it was found that especially p42 and p44 MAP kinases are prominently tyrosine-phosphorylated in response to GM-CSF in a temperature-dependent fashion in vitro, even in non-proliferative neutrophils (McColl et al. 1991; Okuda et al. 1992). The guanylate cyclase system is another possible pathway of GM-CSF-induced signal transduction in neutrophils. GM-CSF raised cGMP levels (Rapoport 1992), an effect that was abrogated by pertussis toxin, which inactivates certain G proteins. Some of the effects of GM-CSF on human PMN were attenuated by a guanylate cyclase inhibitor (Coffey et al. 1993). Lastly, the direct stimulation of arachidonic acid, LTB_4 and PAF (platelet activating factor) synthesis and release may play a role in signal transduction (Rapoport 1992). An inhibitor of 5-lipoxygenase attenuated some of the effects of GM-CSF on human neutrophils (Coffey et al. 1993).

Protein kinase C and other kinases may be involved in down-modulating the GM-CSF receptor and signal, since exposure of neutrophils to protein kinase inhibitors impaired receptor internalization after ligand binding and enhanced the effects of GM-CSF on neutrophils (Rapoport 1992). TNF, PMA and fMLP downregulated GM-CSF receptors, but G-CSF was found to induce increased expression of GM-CSF receptors in a G-CSF responsive, GM-CSF unresponsive murine hematopoietic precursor cell line, resulting in differentiation of the cells in the presence of GM-CSF to monocytes and granulocytes (Rapoport 1992).

1.2 Role in Hematopoiesis

GM-CSF was originally identified as a growth factor capable of supporting the clonal proliferation of granulocyte-macrophage progenitors in culture (reviewed by Clark 1988). It was found that in vitro granulocyte and monocyte progenitor cells die by apoptosis within a few hours in the absence of GM-CSF (Ruef and Coleman 1990; Dexter and Heyworth 1994), whereas they become irreversibly committed to the production of neutrophils, eosinophils and monocytes if the factor is present (Ruef and Coleman 1990). This may be a means of regulating population size of GM-CSF-dependent cells in vivo: There may be a slight overproduction of progenitor cells during steady state, where the excess cells die when the concentration of growth factor becomes limited. In this case some of the additional cells recruited by administration of exogenous GM-CSF would be derived from progenitor cells otherwise destined to die (Dexter and Heyworth 1994). When bipotential granulocyte-macrophage progenitors were treated with high concentrations of GM-CSF in vitro, many cells were forced to enter the granulocytic pathway whereas low concentrations resulted in the formation of pure macrophage progeny (Metcalf 1985). This phenomenon may be explained in part by the observation that GM-CSF downregulates M-CSF-receptor expression (Rapoport 1992). In the presence of M-CSF, suboptimal concentrations of GM-CSF greatly enhanced human macrophage colony-formation (Clark 1988). The synergism between M-CSF and GM-CSF concerning proliferation – here using murine alveolar macrophages – was found to be one-sided: The cells' response to M-CSF was greatly enhanced by the concurrent addition of low doses of GM-CSF, but not vice versa (Chen et al. 1988). Macrophages from GM-CSF cultures were rounder, less stretched and displayed less FcR-mediated phagocytic activity than those produced in M-CSF cultures (Chen et al. 1988). TNF, LT and IFN-γ each suppressed the clonal proliferation of granulocytes, macrophages and mixed granulocyte/macrophage colonies stimulated by GM-CSF (Barber et al. 1987), although GM-CSF-stimulated cells were less sensitive toward these factors than cells stimulated by G-CSF. GM-CSF also markedly enhanced the proliferation of liver macrophages of rats pretreated in vivo with LPS (Feder and Laskin 1994).

GM-CSF can also stimulate megakaryocyte colony-formation and can act in synergy with erythropoetin to further the growth of erythrocyte precursors in vitro (Donahue et al. 1985; Sieff et al. 1985; Ruef and Coleman 1990). GM-CSF was shown to have burst-promoting activity on peripheral blood erythroid progenitors in an in vitro assay involving the delayed addition of erythropoietin (Donahue et al. 1985). It was also shown to support the growth of colony-forming units of blast pluripotential progenitors of granulocytes, erythrocytes, monocytes and macrophages, progenitors of megakaryocytes and blast-forming units of erythrocytes. Furthermore, GM-CSF in combination with IL-3 could stimulate the growth of mast cells, although their numbers were decreased by GM-CSF in the absence of IL-3 (Rottem et al. 1994).

In vitro, GM-CSF supported the growth and function of Langerhans' cells (Ruef and Coleman 1990) and stimulated TRAP$^+$ (tartrate resistent acid phosphatase, a marker for osteoclasts) cells from monkey bone marrow cells expressing the CD34 antigen (Povolny and Lee 1993). GM-CSF is involved in the proliferation and differentiation of murine dendritic cell progenitors to dendritic cells with strong T-cell stimulatory function (Inaba et al. 1992).

An in vivo experiment with rhesus monkeys showed an increase in leukocytosis (granulocytes fivefold, lymphocytes twofold to fourfold, monocytes threefold to fourfold, platelets and erythrocytes unaffected) during treatment with glycosylated or non-glycosylated GM-CSF for seven days. S.c. administration was found to be more effective than i.v. infusion. Within a week after termination of treatment, the WBC had normalized (Mayer et al. 1987). Parallel results were achieved in macaque monkeys (Donahue et al. 1986). In these studies, GM-CSF was also given to one animal for a month, during which time the WBC was maintained at very high levels without any apparent adverse effects.

The leukocytosis initiated by administration of GM-CSF is dose dependent (Scarffe 1991) and phase I studies in humans have found that route and schedule are also important, with lower doses required for s.c. or continuous i.v. administration for responses similar to those achieved with short i.v. infusion of higher doses (Scarffe 1991). In clinical practice, usually no extra production of red cells and platelets was observed, when patients were treated with GM-CSF, so perhaps GM-CSF acts only as a proliferation inducer, but not as a survival stimulus on the precursors of these cells (Aglietta et al. 1989; Dexter and Heyworth 1994). However, there is a study where GM-CSF was used in combination with pentavalent antimony against *Leishmania chagasi*, where an increase in hemoglobin levels was observed at days 5 and 10 of treatment and a rise in platelet count on day 30 (Badaró et al. 1994), suggesting that there might be a delayed effect. Peripheral blood progenitor cells increased after GM-CSF treatment (Scarffe 1991), but little increase in bone marrow progenitors has been observed, because of the dilutional effect of increased marrow cellularity (Scarffe 1991). GM-CSF increased the birth rate of cycling cells and decreased the duration of the S phase and the cell cycle time (Aglietta et al. 1989). The discontinuation of GM-CSF treatment in patients was followed by a period of relative refractoriness of bone marrow cells toward cell cycle-active antineoplastic agents (Aglietta et al. 1989). GM-CSF aided T-cell recovery and was found to favor the regeneration of $CD4^+$ cells after autologous bone marrow transplantation (Miguel et al. 1996).

Many primary myeloid leukemias and leukemic cell lines as well as several T-cell-acute lymphocytic leukemias are dependent on (or responsive to) GM-CSF for growth in culture (Clark 1988). Although there was concern that the hematopoietic growth factors such as GM-CSF might stimulate the growth of solid tumors or leukemias, such stimulation has not been observed to date (Bokemeyer and Schmoll 1995; Hast et al. 1995).

2 Effects on Granulocytes

2.1 Effects on the Functions of Neutrophilic Granulocytes

GM-CSF displays direct and indirect actions on neutrophils. The indirect action is a priming effect which fortifies the cells for enhanced responsiveness to physiologically relevant agents (chemotactic factors, leukotrienes, PAF, IL-8), which may be released locally at sites of inflammation (Di Persio and Abboud 1992). These secondary stimuli can directly activate chemotaxis, the respiratory burst, arachidonic acid metabolism, calcium fluxes, phagocytosis and degranulation, but their effects are dramatically potentiated by pre-exposure to GM-CSF. It seems that the action of GM-CSF on neutrophils is the priming for enhanced functions in a sequence of time-dependent physiologic events corresponding to chemotaxis (5 to 15 min), immobilization (30 min), phagocytosis (1 to 2 h) and enhanced oxidative metabolism (2 h) (Weisbart et al. 1989).

A Phagocytosis

GM-CSF induced a rapid change from low- to high-affinity neutrophil CD89 (IgA Fc receptors), which is associated with IgA-mediated phagocytosis (Weisbart et al. 1988). The phagocytosis of opsonized *Staphylococcus aureus* (Fleischmann et al. 1986) and opsonized *E. coli* (Liehl et al. 1994) by neutrophils, through an increase in the percentage of cells phagocytozing and the number of cells ingested, was stimulated by GM-CSF in vitro, but non-opsonized *S. aureus* were still not ingested (Fleischmann et al. 1986). GM-CSF restored neutrophils' *S. aureus* killing capacity suppressed with dexamethasone (Bober et al. 1995a). The internalization and killing of pathogenic parasites, such as *Trypanosoma cruzi* (Villalta and Kierszenbaum 1986), serum-opsonized yeast (Lopez et al. 1986; Bober et al. 1995a) and *Candida albicans* (Fabian et al. 1992; Bober et al. 1995a) by human PMN was enhanced in the presence of GM-CSF in vitro.

When monkeys were given GM-CSF, their granulocytes displayed enhanced killing of an *E. coli* strain ex vivo (Mayer et al. 1987).

B
Oxidative Burst

Direct triggering of ROS release in suspended neutrophils by GM-CSF is controversial (Yuo et al. 1990; Balazovich et al. 1991). The direct GM-CSF-induced release of ROS that was reported could be inhibited by cyclic AMP agonists (PGE_1 and dibutyryl cAMP) or cytochalasin B (Yuo et al. 1990).

PMN were primed by GM-CSF to increase the production of ROS in response to the bacterial chemoattractant fMLP (Weisbart et al. 1985; Lopez et al. 1986; Khwaja et al. 1992; Treweeke et al. 1994) in vitro and ex vivo (Sullivan et al. 1989a; Kaplan et al. 1989; Khwaja et al. 1992) as well as in vitro to C5a (Weisbart et al. 1987; Khwaja et al. 1992), α-hemolysin (Konig and Konig 1994) or LTB_4 (Weisbart et al. 1987), but not in response to PMA (Weisbart et al. 1987). Augmented oxidative response was recorded towards pathogenic parasites (Villalta and Kierszenbaum 1986) in vitro and toward bacteria, here *E. coli* in monkeys, in vivo (Mayer et al. 1987). In another study, cytochalasin B was required in addition to fMLP to enhance superoxide production of GM-CSF-primed neutrophils in vitro (Nagata et al. 1995). Here, there was no inhibition with anti-CD18 antibody, but instead with a 5-lipoxygenase-activating protein antagonist. The CD32 (FcγRII)-mediated production of superoxide was augmented by GM-CSF without a change in CD32 expression in vitro in studies of neutrophils from healthy individuals or in vivo in studies of patients receiving GM-CSF (Roberts et al. 1990). In summary, GM-CSF appears to prime receptor-mediated initiation of the oxidative burst rather than PMA-induced ROS formation.

The enhanced neutrophil oxidative response to stimulation with fMLP after priming with GM-CSF was induced in a temperature-dependent manner (potentiated at 37 °C) and extracellular Ca^{2+} was not required for functional enhancement. Also, there was no alteration of the phospholipid content effected by incubation in vitro with GM-CSF alone (English et al. 1988). The acceleration in the rate of ROS release was accompanied by an antecedent increase in membrane depolarization, which was independent both of the resting transmembrane potential and of alterations in the extent of membrane potential change induced by stimuli such as fMLP (Sullivan et al. 1989b).

Human neutrophils adherent to proteins derived from serum or plasma, or to laminin showed a markedly more delayed, prolonged and greater respiratory burst in response to soluble agonists when primed by GM-CSF in comparison to neutrophils in suspension (Nathan 1989).

The addition of monocytes and lymphocytes to PMN resulted in a near doubling of GM-CSF-primed fMLP-stimulated ROS release by the PMN. No cell-free "enhancing factors" could be detected, but cell-to-cell contact further enhanced this oxidative activity. Polyclonal rabbit anti-TNF antibody decreased the extent of the oxidative burst by fMLP-stimulated GM-CSF-primed PMN as well as by the leukocyte mixture, suggesting that TNF on the PMN surface might enhance GM-CSF-primed oxidative burst (Sullivan et al. 1993).

C
Adhesion, Chemotaxis and Migration

Although one study states that GM-CSF did not influence the adhesion of neutrophils to plastic surfaces or endothelial cells (Lopez et al. 1986), other studies found that GM-CSF increased adherence of PMN to plastic surfaces (Zeck-Kapp et al. 1989) and increased neutrophil adhesion to cultured human endothelium in vitro by upregulating CD11b (Yong and Linch 1992) and CD11c and downregulating CD62L (Yong and Linch 1992). GM-CSF also enhanced cell-to-cell adhesion of human mature granulocytes (Arnaout et al. 1986), an effect which was inhibited by a monoclonal antibody directed against the CD11b antigen. Furthermore, GM-CSF potentiated PMN aggregation responses to fMLP, but did not induce this response when administered alone (Conti et al. 1992).

There are conflicting statements on whether GM-CSF is itself a chemotactic factor (Wang et al. 1987) or merely chemokinetic (Yong and Linch 1993; Smith et al. 1994). In the latter two studies, the presence of GM-CSF enhanced migration of neutrophils across filters or unstimulated endothelium in vitro dose-dependently, but independent of its own concentration gradient and to a lesser extent than with IL-8. In contrast, GM-CSF inhibited neutrophil migration across IL-1-activated endothelium and almost completely abolished IL-1-induced migration (Yong and Linch 1993). Similarly, exposure of neutrophils to GM-CSF decreased their migration through TNF-activated endothelial monolayers (Smith et al. 1994), suggesting little role for GM-CSF in neutrophil diapedesis at inflammatory sites in vivo.

GM-CSF had a biphasic effect on neutrophil motility: within 5 to 15 min after incubation with GM-CSF, there was enhanced chemotaxis along a gradient to fMLP, a potent chemoattractant (Weisbart et al. 1986; Bober et al. 1995a). Further experiments showed that enhanced chemotactic responses to fMLP were no longer evident after prolonged incubation of neutrophils with GM-CSF (Weisbart et al. 1986; Kharazmi et al. 1988). The migratory capacity of neutrophils toward LTB_4 was also increased by GM-CSF (Bober et al. 1995a). However, an ex vivo study using PMN from carcinoma patients treated with GM-CSF, could not find enhancement of chemotaxis in response to fMLP or C5a (Kaplan et al. 1989).

Furthermore, GM-CSF inhibited migration and locomotion of neutrophils to sterile inflammatory sites (Peters et al. 1988). After incubation with GM-CSF for 30 to 120 min there was a decrease in "random migration" of neutrophils (Di Persio and Abboud 1992). These observations were confirmed in vitro (Di Persio and Abboud 1992) and in vivo using a micropore skin window technique (Addison et al. 1989), suggesting that GM-CSF may impair the ability of neutrophils to infiltrate an inflammatory focus.

In mice with an inactivated GM-CSF gene, typical neutrophil migration to an inflammatory site still took place, confirming that GM-CSF plays no major role in this mechanism (Metcalf et al. 1996). Taken together, only a modest or no augmentation of leukocyte migration can be attributed to GM-CSF.

D
Tumor Cytotoxicity

Purified GM-CSF enhanced the ADCC of human neutrophils toward various tumor targets; this stimulation was rapid in onset and required direct contact with the targets for killing (Vadas et al. 1983). In another study, both purified normal mouse bone marrow neutrophils or induced peritoneal neutrophils displayed enhanced ADCC activity when incubated with rmuGM-CSF and could kill TNP-coupled mouse thymoma cells coated with anti-TNP antibodies at low antibody concentrations. RmuGM-CSF and rmuG-CSF had an additive effect in this case (Lopez et al. 1983). RhuGM-CSF potentiated 3F8-mediated ADCC of G_{D2}-positive tumor targets (melanoma and neuroblastoma) by human granulocytes in vitro, whether GM-CSF was present in the ADCC assay or whether granulocytes were incubated with GM-CSF and washed before the assay (Kushner and Cheung 1989). Non-oxidative mechanisms may be important for ADCC, as this phenomenon was also observed with granulocytes from two children with chronic granulomatous disease (Kushner and Cheung 1989). In line with this finding, the GM-CSF-enhanced ADCC activity toward neuroectodermal tumor target cells can be inhibited in the presence of antibody to CD32 (Baldwin et al. 1993).

E
Synthesis of Mediators and Enzymes

Human neutrophils were primed to produce and release arachidonic acid by GM-CSF in response to the chemotactic factors fMLP, LTB_4, PAF as well to calcium ionophore in a concentration- and time-dependent manner (Sullivan et al. 1987; Di Persio et al. 1988c). The prompt release of arachidonic acid from plasma membrane phospholipids is an event which may represent the receptor-mediated activation of membrane phospholipases and that may contribute to the priming of the cells for enhancement of their functional responsiveness (Sullivan et al. 1987). However, it was determined in another study that GM-CSF is unable to induce the release of cell-incorporated arachidonic acid or to increase the level of phosphatidic acid directly (Ulich et al. 1990a).

PMN were also primed by GM-CSF to synthesize the arachidonic acid metabolite LTB_4 and its derivatives when induced by fMLP and cytochalasin (Nagata et al. 1995) or C5a (Dahinden et al. 1988), fMLP (Di Persio et al. 1988b; Dahinden et al. 1988), the ionophore A23187 (Di Persio et al. 1988b and c; Hensler et al. 1994), α-hemolysin (Konig and Konig 1994), PAF (McColl et al. 1991) or leukocidin from *Staphylococcus* (Hensler et al. 1994) in a dose-, time- and temperature-dependent manner (Di Persio et al. 1988b), though GM-CSF alone has no effect on LTB_4 production (Di Persio et al. 1988b; Conti et al. 1992). The enhancing effect of GM-CSF was ablated when neutrophils were stimulated with ionophore and exogenous arachidonic acid, but co-addition of arachidonic acid with fMLP did not entirely mask the effect of GM-CSF (Di Persio et al. 1988b). LTB_4 produces the same enhancement of the oxidative burst function as does the incubation with GM-CSF, suggesting that LTB_4, induced by GM-CSF, may transmit the main effect on superoxide production (Nagata et al. 1995) and may also play a role in the promotion of neutrophil chemotaxis by modulating phospholipid methylation (Di Persio and Abboud 1992).

GM-CSF further primed neutrophils for increased PAF synthesis in response to secondary stimuli, but the importance of cell-associated PAF remains controversial. Arguments are reviewed by Di Persio and Abboud 1992.

IL-6 (Cicco et al. 1990; Sallerfors 1994) and also IL-8, a potent chemotactic factor for neutrophils(Takahashi et al. 1993; Gatti et al. 1995), M-CSF, G-CSF and TNF-α (Lindemann et al. 1989b) were produced by neutrophils on stimulation with GM-CSF in vitro. However, it seems that IL-8 release from human PMN challenged with *E. coli* α- hemolysin was not affected by the addition of GM-CSF (Konig and Konig 1994). The levels of IL-1ra mRNA and protein synthesis were increased transiently by GM-CSF though an increased transcription of IL-1β was detected in some studies, but not in others (Lindemann et al. 1988; Re et al. 1993; Malyak et al. 1994; Fernandez et al. 1996). PMN may therefore be a major source of IL-1ra, which has been shown to inhibit the in vitro and in vivo effects of IL-1 in inflammatory exsudates where these cells predominate. Neutrophils isolated from an inflammatory milieu, i.e. the synovial fluid of patients with rheumatoid arthritis, were found to respond to GM-CSF in terms of IL-1ra synthesis, indicating that the in vitro observations are likely to occur in an inflammatory setting in vivo (McColl et al. 1992).

IL-8 release, but not IL-6 or TNF-α, was induced in healthy volunteers injected with a single dose of GM-CSF (van Pelt et al. 1996). The levels of IL-8 and IL-6 in the bronchoalveolar lavage fluid of patients with unresectable small-cell lung cancer treated with GM-CSF were increased shortly after the treatment in concordance with the increased neutrophil and macrophage levels (Gatti et al. 1995).

In summary, GM-CSF augments the formation and release of a number of inflammatory mediators such as eicosanoids, PAF and cytokines by PMN.

F
Degranulation

Degranulation was induced by GM-CSF directly in a dose-dependent manner as assessed by measurement of the cytochalasin B-induced release of MPO from primary granules and lactoferrin from secondary granules (Richter et al. 1989; Treweeke et al. 1994). This effect is possibly mediated via a GTP- binding protein and/or changes in local intracellular calcium concentrations, as it was partly prevented by pertussis toxin and an intracellular calcium buffer. GM-CSF also stimulated the degranulation and the release of the enzymes β-glucuronidase from primary and arylsulphatase from secondary granules (Fabian et al. 1992) and induced enhanced degranulation of vitamin B_{12}-binding protein from secondary granules that can be increased further by the addition of fMLP, PAF or ionophore A23187 (Kaufman et al. 1989), although another study found that the release without prior incubation with cytochalasin B is negligible and also found no MPO in the supernatant (Smith et al. 1990c).

GM-CSF enhanced the degranulation of fMLP-stimulated, cytochalasin B pretreated neutrophils as measured by the release of lysozyme (Lopez et al.

1986; Coffey et al. 1993) and β-glucuronidase (Coffey et al. 1993). However, the release of β-glucuronidase from human PMN challenged with α-hemolysin was not affected by the addition of GM-CSF (Konig and Konig 1994) and GM-CSF does not affect the NAP synthesis in vitro, in spite of enhanced incorporation of amino acids into PMN, but GM-CSF suppresses the enhancement of both of these functions by G-CSF (Teshima et al. 1990).

In an in vivo study with healthy volunteers, a single injection of GM-CSF resulted in increased serum levels of the degranulation products lactoferrin and elastase (van Pelt et al. 1996).

G
Expression of Surface Molecules

Since GM-CSF can directly induce degranulation and since both fMLP receptors and CD11b are stored in specific granules, enhanced degranulation may have a major role in the increased surface expression of both of these proteins (Di Persio 1990). GM-CSF also displays acute upregulatory effects on PMN by mobilizing CD35 and CD18/CD11 complexes from intracellular compartments within 30 min, a process which does not depend on protein synthesis (Neuman et al. 1990).

In vitro, the expression of the PMN surface antigens CD11b, CD14, CD18/CD11c, CD35, CD54 and CD64 was increased in the presence of GM-CSF (Lopez et al. 1986; Arnaout et al. 1986; Griffin et al. 1990; Neuman et al. 1990; Treweeke et al. 1994; Bober et al. 1995a; Yong 1996; Kruger et al. 1996a). GM-CSF increased the number and affinity of neutrophil CD89 (Weisbart et al. 1988) and down-regulated the expression of IL-6 receptors (Henschler et al. 1991), LTB_4 receptors and IL-8 receptors (Di Persio and Abboud 1992). During short incubation times with GM-CSF, there was a rapid increase in the number of high-affinity fMLP receptors expressed on PMN; a more prolonged incubation was accompanied by a change to low-affinity fMLP receptors (Weisbart et al. 1986; Yuo et al. 1990; Gasson 1991). The leukocyte adhesion molecule CD62L, which mediates the binding of leukocytes to human high endothelial venules of peripheral lymph nodes and of neutrophils to the endothelium at inflammatory sites, was seen to be downregulated in the presence of GM-CSF in vitro (Griffin et al. 1990; Spertini et al. 1991; Yong 1996).

GM-CSF can induce MHC class II antigens in pure cultures of PMN (Gosselin et al. 1993), though this induction was found to be distinctly donor dependent. Their potential to express class II antigen suggests that PMN could play a significant role in immunoregulation and disease pathogenesis (Gosselin et al. 1993). Although freshly isolated peripheral blood neutrophils neither bind nor respond to IL-3, incubation with GM-CSF resulted in the expression and transcription of the IL-3 receptor α-subunit (Smith et al. 1995). Addition of IL-3 to the GM-CSF medium led to an increase in HLA-DR expression on PMN that was greater than with GM-CSF alone (Smith et al. 1995). GM-CSF also induced the synthesis of MHC class I protein (Neuman et al. 1990).

In vivo, administration of GM-CSF resulted in changes that were similar although of lesser magnitude (Dale 1994). An in vivo study on patients undergoing GM-CSF treatment confirmed the in vitro observations that CD11b and CD35 expression by neutrophilic granulocytes is markedly increased, while there was a substantial decrease or even loss of CD16 (Socinski et al. 1988; Maurer et al. 1991). These results were also documented in a study with healthy volunteers in addition to a decrease in CD62L on neutrophils (Kishimoto et al. 1996). CD11a and CD11c expression remains unchanged (Socinski et al. 1988). Another study confirms that there is a rapid in vivo increase in cellular adhesion molecule (CAM) expression after GM-CSF administration, paralleling the development of transient neutropenia due to margination in the pulmonary vasculature. These data suggest that adhesion promoting glycoproteins play a part in margination (Devereux et al. 1989). In a patient with partial leukocyte adhesion deficiency, GM-CSF administration in vivo was still associated with margination of neutrophils despite no detectable upregulation of CD11b, implying other mechanisms in this process (Di Persio and Abboud 1992). Demargination occured at a time when neutrophil CAM expression was still high, so the dissociation of the neutrophil-endothelial cell interaction is likely to depend on other factors (Devereux et al. 1989).

These rapid and direct effects on surface adhesion molecule or receptor expression may result in a more efficient migration of mature and primed effector neutrophils to areas of inflammation by preventing extravasation into and damage of normal tissues by activated neutrophils (Di Persio 1990; Di Persio and Abboud 1992). However, the importance of this effect in vivo is questionable (see chemotaxis).

H
Other

The survival of neutrophils was increased in vitro by GM-CSF which protects the cells from apoptosis, i.e. programmed cell death (Lopez et al. 1986; Begley et al. 1986; Colotta et al. 1992; Brach et al. 1992).

GM-CSF induced morphological changes in human PMN in vitro such as the extrusion of filamentous filopodia, polarization and the development of intracellular vesicles (Zeck-Kapp et al. 1989; Coffey 1989). GM-CSF seems to have a direct, but slow effect on neutrophil F-actin polymerization as seen in vitro, and also primes neutrophils for F-actin polymerization. This process may increase the mechanical stiffness and neutrophil trapping in the microvasculature of the lungs (Di Persio and Abboud 1992).

The cytotoxic function of neutrophils from patients with AIDS towards HIV-infected MOLT-3A cells was markedly augmented by GM-CSF in vitro (Baldwin et al. 1989).

2.2 Effects on the Functions of Eosinophilic Granulocytes

A Phagocytosis

GM-CSF enhanced the phagocytosis of *Candida albicans* by human eosinophils in vitro (Fabian et al. 1992), although their killing capacity was not affected.

B Oxidative Burst

FMLP-stimulated superoxide generation by eosinophils was significantly enhanced when incubated in the presence of GM-CSF (Nagata et al. 1995; Sedgwick et al. 1995). This was inhibited by an anti-β_2 (CD18) antibody in the same study, suggesting that CD18 is associated with the increased cell function (Nagata et al. 1995).

C Adherence, Chemotaxis and Migration

GM-CSF increased the adhesion of eosinophils to tissue culture plates (Nagata et al. 1995) and increased the PAF- or fMLP-induced adherence to gelatin-coated plastic (Tomioka et al. 1993). Spontaneous adhesion to human umbilical vein endothelial cell monolayers was increased after exposure of human eosinophils to GM-CSF (Sedgwick et al. 1995). Furthermore, preincubation of eosinophils with GM-CSF their mean adhesion strength to endothelial cells mediated by the VLA-4 (very late antigen 4) receptor, increased significantly through the transition of VLA-4 from low to high affinity (Sung et al. 1997).

Pre-incubation of eosinophils with picomolar concentrations of GM-CSF caused a significant increase in the chemotactic response toward LTB_4 and induced a significant chemotactic response toward IL-8 and fMLP. Nanomolar concentrations of GM-CSF inhibited C5a-induced as well as LTB_4-induced chemotaxis. If, on the other hand, the cells were washed after preincubation with GM-CSF the potentiation of the chemotactic response remained and the inhibitory action disappeared (Warringa et al. 1991).

Intraperitoneal injections of GM-CSF into guinea pigs and subsequent exposure to an aerosol of PAF resulted in a selective pulmonary eosinophil accumulation which was significantly enhanced in comparison to control animals that received no CSF or which were injected with IL-3 (Sanjar et al. 1990).

D
Cytotoxicity

GM-CSF increased the cytotoxicity of eosinophils toward various target cells (Vadas et al. 1983; Lopez et al. 1986; Owen et al. 1987), such as a number of tumors, antibody coated targets or toward *Schistosoma mansoni* larvae by enhanced ADCC in vitro (Silberstein et al. 1986). In the latter reaction, not only the eosinophil schistosomicidal ability is increased, but the threshold for antibody or complement required in the killing reaction was also lowered (Dessein et al. 1982): antibody-coated larvae were frequently found covered by several layers of eosinophils in tubes containing GM-CSF. GM-CSF also enhanced the temperature-dependent reaction that ensures the irreversibility of eosinophil attachment to schistosomula (Dessein et al. 1982).

E
Synthesis of Mediators

Calcium ionophore A23187-induced generation of leukotriene C_4 (LTC_4) was augmented in vitro (Silberstein et al. 1986; Fabian et al. 1992; Nagata et al. 1995) and IL-2 mRNA expression in human eosinophils increased upon costimulation with ionophore and GM-CSF (Bosse et al. 1996).

F
Degranulation

GM-CSF activated eosinophils to transform the storage form of eosinophil cationic protein into the secreted form and enhanced its secretion when induced by a secretory signal, such as sepharose coated with C3b or sepharose-activated whole autologous serum (Tai et al. 1990). The release of arylsulphatase and β-glucuronidase from granules of human eosinophils was induced by GM-CSF (Fabian et al. 1992).

G
Expression of Surface Molecules

Incubation of murine eosinophils with GM-CSF produced a potent induction of the transcription of CD32 (FcγRII) and CD16 (FcγRIII) (de Andres et al. 1994) and increased membrane expression of PAF receptors (Kishimoto et al. 1996), CD18 as well as CD11b (Sedgwick et al. 1995; Tomioka et al. 1993) in comparison to the constitutive expression in cultures without GM-CSF. In addition, in vitro incubation of human eosinophils with GM-CSF resulted in and induced increased CD11b responses to PAF and fMLP (Tomioka et al. 1993). After a single in vitro injection of GM-CSF into healthy volunteers their eosinophils also displayed an upregulation of CD11b (van Pelt et al. 1996).

H
Microbial Killing

The killing of *Staphylococcus aureus* by human eosinophils was increased by GM-CSF, but not that of *Candida albicans* (Fabian et al. 1992).

I
Other

Eosinophils survived longer in vitro in the presence of GM-CSF (Lopez et al. 1986; Begley et al. 1986; Tai et al. 1990 and 1991). The in vitro survival of peripheral blood eosinophils without CSF is limited to two days before they die by apoptosis (Tai et al. 1991). Of normo-dense human peripheral blood eosinophils cultured with GM-CSF, 43% survived for seven days and in the additional presence of mouse 3T3 fibroblasts they survived for at least 14d. The resulting eosinophils became hypo-dense and had an augmented capacity to generate LTC_4 and to kill *Schistosomula mansoni* larvae (Owen et al. 1987). The prolongation of eosinophil survival is inhibited by the glucocorticoids dexamethasone, prednisolone and hydrocortisone by signals mediated via a glucocorticoid receptor (Lamas et al. 1991). The sex steroids testosterone and β-estradiol had no such effect. Culture of eosinophils with cholera toxin produced a concentration-dependent decrease in GM-CSF-induced survival that was associated with an increase in the cAMP concentration (Hallsworth et al. 1996). This inhibition of cell survival could be prevented by the addition of a protein kinase inhibitor. GM-CSF-induced cell survival could also be prevented with dibutyryl cAMP, but not with dibutyryl cGMP (Furman et al. 1992). The survival stimulus transmitted by GM-CSF apparently does not affect intracellular cAMP levels (Hallsworth et al. 1996).

2.3 Effects on the Functions of Basophilic Granulocytes

A Histamine Release

One study found that the direct release of histamine by basophils of most healthy volunteers was increased in the presence of GM-CSF (Haak-Frendscho et al. 1988), while another found that only relatively high doses of GM-CSF caused a release of small amounts of histamine from the cells of few allergic donors, not from cells of non-allergic donors (Alam et al. 1989). Histamine release could be achieved with lower concentrations of GM-CSF in the presence of D_2O by some allergic donors' cells (Alam et al. 1989).

GM-CSF markedly enhanced histamine release in a dose dependent manner upon stimulation with anti-IgE, fMLP or ionophore (Hirai et al. 1988; Miadonna et al. 1993). The enhancement was rapid and temperature dependent, but no additive effect was observed when GM-CSF and IL-3 were combined (Hirai et al. 1988). GM-CSF was found to render basophils capable of responding to C3a by releasing histamine, but the effect was smaller than with IL-3 (Bischoff et al. 1990).

B Chemotaxis

GM-CSF had potent human basophil chemotactic activity at picomolar concentrations, but whether it primarily induces chemotaxis or simply chemokinesis is controversial (Tanimoto et al. 1992; Yamaguchi et al. 1992a).

C Other

GM-CSF had a slight positive effect on the survival of human basophils in culture (Yamaguchi et al. 1992b).

3
Effects on Mononuclear Cells

3.1
Effects on the Functions of Monocytes/Macrophages

Incubation of murine peritoneal macrophages with GM-CSF induced a rapid spreading and increase in cell size within 24h, suggesting a stimulation of metabolism: an increase in RNA and protein synthesis was detected, but DNA synthesis was not induced (Heidenreich et al. 1989).

A
Maturation of Monocytes to Macrophages

GM-CSF stimulated the differentiation of human blood monocytes to macrophages under various in vitro conditions with a higher yield than M-CSF or IL-3 (Lopez et al. 1993). The cells displayed similar CD14, CD64, CD71, HLA-DR and Max1 antigen expression and similar in vitro anti-tumoricidal activity against U937 cells, whether the differentiation was stimulated with GM-CSF, M-CSF or IL-3 or simply in culture with autologous lymphocytes. Macrophages derived from monocytes through incubation with GM-CSF in culture secreted more TNF in response to LPS than freshly isolated monocytes (Ross and Koeffler 1992) which suggests that GM-CSF may influence TNF production by accelerating the maturation of monocytes to macrophages.

B
Phagocytosis

Although GM-CSF was found to stimulate the phagocytosis of *Cryptococcus neoformans*, an encapsulated fungal pathogen that was opsonized with complement, by peritoneal macrophages in vitro. This effect was dramatically up-regulated when GM-CSF and TNF-α synergized at low concentrations, such that the anti-phagocytic properties of the virulent yeast were overcome completely (Collins et al. 1992). Elutriated human monocytes primed with GM-CSF reduced the number of viable *Candida albicans* in culture compared with non-primed monocytes and monocytes incubated with G-CSF (Smith et al. 1990a; Liehl et al. 1994). Pretreatment of murine macrophage monolayers with GM-CSF before infection led to an increased level of phagocytosis of *Leishmania tropica* (Handman and Burgess 1979). Treatment of already infected murine monolayers with GM-CSF caused the intracellular parasites to undergo morphologic damage whereas they survive unharmed in control macrophages (Handman and Burgess 1979). GM-CSF stimulated Fc-dependent phagocytosis by peritoneal macrophages, although the stimulation of resident macrophages was less dramatic than that of thioglycolate-elicited cells (Coleman et al. 1988).

In an experiment where mice were injected with rmuGM-CSF for a period of 11 weeks, the number and phagocytic activity of peritoneal macrophages toward Indian ink droplets, which increased in the beginning, normalized despite the continuing GM-CSF treatment as did other hematological parameters (Podja et al. 1989).

An example of enhanced phagocytosis in patients was the rapid fall in platelet counts in a patient with auto-immune thrombocytopenia, most likely due to enhanced removal of partially damaged platelets (Jones 1993).

C
Oxidative Burst

GM-CSF does not affect the respiratory burst of adherent monocytes (Nathan 1989), but instead does affect that of monocytes in suspension by priming them and enhancing the release of ROS stimulated by fMLP or ConA, but not by PMA which bypasses the receptors to stimulate the cells (Yuo et al. 1992). The priming effect of GM-CSF was greater than that of M-CSF or IL-3 and was achieved in only 10 min of pre-incubation. Incubation with GM-CSF induces an incremented monocyte oxidative response to both an anti-CD32-antibody and fMLP, although the apparent increase in priming is less than that seen in neutrophils, although this may depend on the setup of the experiment (Roberts et al. 1990). Another study states that rmuGM-CSF increased PMA or opsonized zymosan-elicited H_2O_2 release by resident and thioglycolate-elicited murine macrophages after 48h in vitro (Coleman et al. 1988).

Monocytes from patients treated with a high dose of GM-CSF over a period of 14 d displayed no increase in basal superoxide anion production ex vivo, but their production capacity when maximally stimulated with phorbol dibutyrate increased, though not significantly (Perkins et al. 1993). In another study, the priming effect observed in vitro when monocytes from patients who had undergone high dose chemotherapy were stimulated with LPS or opsonized *Staphylococcus aureus* even lasted several weeks after the cessation of GM-CSF therapy (Williams et al. 1995).

D
Adhesion, Chemotaxis and Migration

Monocytes showed an enhanced adhesion to plastic and to endothelial cells in vitro in the presence of GM-CSF (Gamble et al. 1989). GM-CSF also induced polarization and migration of human peripheral blood monocytes (Wang et al. 1987).

E
Tumor Cytotoxicity

GM-CSF stimulated peripheral blood monocytes in vitro to become cytotoxic towards a malignant melanoma cell line (Grabstein et al. 1986a) and two ovarian tumor cell lines (Bernasconi et al. 1995) without further exogenous signals. This activity was also observed with murine peritoneal macrophages in vitro towards TNF-α insensitive Eb lymphoma cells (Heidenreich et al. 1989). Conditioned medium containing GM-CSF, however, failed to activate macrophages for effector activities against fibrosarcoma, lymphoma or *Leishmania tropica* amastigotes (either resistance to infection or intracellular destruction) (Ralph et al. 1983). GM-CSF treatment of freshly isolated monocytes resulted in enhanced ADCC against melanoma target cells, especially when GM-CSF was used in conjunction with a secondary stimulus (Baldwin et al. 1993). Cytotoxicity of peripheral blood monocytes towards some other tumor cell lines also requires an additional exogenous signal to GM-CSF, such as LPS. This was the case in a study employing WEHI 164 fibrosarcoma cells as targets (Cannistra et al. 1988), where it was suggested that the augmented activity was mediated through increased release of TNF, as the cytokine-enhanced effect was abolished by anti-TNF antibody. Murine peritoneal macrophages failed to generate cytotoxicity toward TNF-α sensitive L929 cells when treated with GM-CSF alone, but were primed by GM-CSF to produce increased levels of TNF-α in response to IFN-γ plus LPS (Heidenreich et al. 1989). This priming effect of GM-CSF disappeared upon longer incubation (>12 h) and was followed by reduced responsiveness to signals which induce TNF-α release.

A comparison between the dose-response curves of alveolar macrophages and monocytes towards GM-CSF showed dramatic differences in their cytotoxic activity: alveolar macrophages exhibited significant cytotoxic activity at all doses tested, while monocytes displayed significantly less activity than macrophages at every dose. Tumoricidal activity seems to be related to maturity, as monocytes matured in vitro showed enhanced activity when exposed to GM-CSF. This tumoricidal activity was not dependent on oxidative metabolism, TNF-α or IL-1β (Thomassen et al. 1989). GM-CSF enhances Kupffer-cell mediated cytotoxicity toward the SW948 colon carcinoma cell line and the tumor line U937 in vitro, an effect which is enhanced by the addition of IFN-γ (Schuurman et al. 1994). TNF-α secretion by human Kupffer cells increased in parallel with their cytotoxicity after incubation with GM-CSF and was identified as the main cytolytic mechanism of human Kupffer-cell-mediated cytotoxicity.

Monocytes derived from GM-CSF-treated patients produced more TNF than did cells from control patients and they showed a significant increase in cytotoxic activity against U937 tumor cells (Wiltschke et al. 1995). In patients with solid malignancies, treatment with GM-CSF led to a significant enhancement of direct monocyte cytotoxicity against the human colon carcinoma (HT29) cell line ex vivo, but there was no increase in serum TNF-α or IL-1β and no consistent in vitro induction of TNF-α or IL-1β from monocytes posttreatment (Chachoua et al. 1994). In a study where patients were treated with GM-CSF i.v., the tumorilytic properties of monocytes were only enhanced significantly in one of seven patients and neither IL-1 nor TNF production were stimulated (Kleinerman et al. 1988). The tumoricidal activity of monocytes from volunteers treated with GM-CSF for three days showed significant elevation when stimulated in vitro with LPS, but the activity of cells harvested after ten days of treatment returned to pretreatment values (Thomassen et al. 1991).

In a murine melanoma model, in which irradiated tumor cells alone do not stimulate significant anti-tumor immunity, irradiated tumor cells engineered to express murine GM-CSF stimulated potent, long-lasting and specific anti-tumor immunity, requiring both $CD4^+$ and $CD8^+$ cells (Dranoff et al. 1993). GM-CSF was also found to be active in this way in other tumor models, such as cell lines derived from colon, renal cell and lung carcinoma as well as fibrosarcoma (Dranoff et al. 1993).

F
Synthesis of Mediators and Enzymes

Elutriated, cultured monocytes responded to GM-CSF with the production of pro-inflammatory IL-1β, IL-6, IL-8 and TNF-α mRNA as well as expression of autoregulatory IL-1ra and the M-CSF gene (Cluitmans et al. 1993; Takahashi et al. 1993). Expression of the IL-8 gene (Takahashi et al. 1993) and of the TNF gene (Cannistra et al. 1987) was also measured in the U937 cell line in response to addition of GM-CSF to the medium. However, the production of IL-1, IL-6, IL-8,TNF protein and PGE_2 was hardly influenced by GM-CSF in blood monocytes in vitro (Hart et al. 1988; Bernasconi et al.

1995). However, in another study, IL-8 secretion by monocytes was stimulated after incubation with GM-CSF (Takahashi et al. 1993), although here GM-CSF also did not induce detectable secretion of IL-1, TNF-α or IL-6 protein by monocytes (Takahashi et al. 1993). Instead, GM-CSF did directly induce the production of cell-associated IL-1 (Danis et al. 1991) and IL-1ra was found in the cell supernatants of adherent peripheral blood monocytes induced by GM-CSF (Shields et al. 1990).

GM-CSF induced the transcription and secretion of M-CSF in monocytes (Horiguchi et al. 1987; Gruber and Gerrard 1992). This effect was enhanced further by TNF-α, IFN-γ or M-CSF (Gruber and Gerrard 1992). Therefore, GM-CSF may indirectly control specific monocyte functions by regulating the production of M-CSF (Horiguchi et al. 1987). Further, GM-CSF alone or in synergistic concert with IL-3 induced the transcription of the G-CSF gene and the release of biologically active G-CSF in vitro from peripheral blood monocytes (Oster et al. 1989b; Sallerfors and Olofsson 1992). Monocytes of human newborn infants did not produce IFN-γ, but have been shown to commence its production in a medium conditioned with GM-CSF in vitro (McKenzie et al. 1993). GM-CSF also enhanced the secretion of IFN-γ by normal monocytes (De Witte et al. 1995).

Combinations of GM-CSF with IFN-γ, IL-2 or TNF-α synergistically enhanced IL-1 secretion and had an additive effect on cell-associated IL-1 production by monocytes (Hart et al. 1988; Danis et al. 1991). GM-CSF had a priming effect on TNF production in response to LPS. Sisson and Dinarello (1988) found that human peripheral blood mononuclear cells in vitro responded to a low dose of GM-CSF by increasing the production of TNF, whereas some donor's cells also produced more IL-1α and/or IL-1β. Indomethacin increased the amounts of TNF and IL-1β produced in response to GM-CSF, but did not affect the values of IL-1α. The primed state was quickly followed by a period of relative unresponsiveness to LPS. This is apparently due to the secretion of PGE_2 in response to GM-CSF, which did not affect the production of TNF mRNA but rather blocked its translation, possibly by a mechanism involving cAMP. This may be one of the mechanisms used in vivo to contain the inflammatory response and tissue destruction. The priming effect was restored by the cyclooxygenase blocker indomethacin (Heidenreich et al. 1989). The temporally delayed generation of TNF-α and subsequently PGE_2 suggests an autoregulatory circuit in which PGE_2 limits GM-CSF-induced macrophage activation.

GM-CSF provided a priming stimulus similar to IFN-γ for LPS-induced TNF and IL-12 p40 mRNA in human monocytes, but primed poorly for the other LPS-inducible IL-12 p35 (Hayes et al. 1995). Furthermore, GM-CSF synergistically enhanced TNF-α secretion induced by IFN-γ but not by LPS,

IL-2 or TNF-α (Hart et al. 1988; Danis et al. 1991; Kohn et al. 1992). GM-CSF potently stimulated IL-8 secretion in cultures of heparinized whole blood (Takahashi et al. 1993), which is primarily released by monocytes under these conditions.

GM-CSF also appears to modulate factor release by mature macrophages. The expression of membrane-bound IL-1 was stimulated and the de novo synthesis of Ia molecules of the murine equivalent of MHC class II was induced by GM-CSF in murine bone marrow-derived macrophages in vitro, but to a lesser extent than with IFN-γ (Fischer et al. 1988). Further, the production of IL-6 and IL-8, but not IL-1 and TNF, was increased in tumor-associated macrophages in the presence of GM-CSF (Bernasconi et al. 1995). The down-regulation of apolipoprotein E secretion by macrophages in response to GM-CSF was found to be mediated by a TNF- dependent autocrine mechanism, as it was inhibited by a monoclonal antibody against TNF (Zuckerman and O'Neal 1994). The promyelocytic cell lines H-161, AML-193 and HL-60, differentiated to adherent macrophage-like cells under the influence of PMA, responded to GM-CSF with the production of IL-1ra (Mazzei et al. 1990).

GM-CSF activated PGE_2 production with IFN-γ, but not with LPS (Hart et al. 1988). The basal production of complement factor C3, but not factor B, by human monocytes was inhibited by GM-CSF (Høgåsen et al. 1993). The LPS-stimulated production of both the complement factors C3 and B was abrogated by GM-CSF, and, in line with this finding, anti-GM-CSF added to unstimulated or LPS- stimulated cells caused an increase in C3 production (Høgåsen et al. 1993). Plasminogen-activator inhibitor PAI-1 and PAI-2 mRNA and protein levels were enhanced by GM-CSF in vitro (Hamilton et al. 1993). They may play a role in modulating the effects of the CSF on monocyte u-PA activity at sites of inflammation and tissue remodeling (Hamilton et al. 1993). However, GM-CSF also seemed to directly stimulate the production of a u-PA in both murine bone marrow-derived and peritoneal macrophages (Hamilton et al. 1991). This increase in u-PA activity was abrogated by dexamethasone, PGE_2 and cholera toxin.

In animal studies, we found GM-CSF to be a potent enhancer of LPS-induced TNF-α production in normal as well as in experimentally immunocompromised (LPS-tolerant) mice (Bundschuh et al. 1997; Randow et al. 1997). Our results in murine endotoxic shock as well as in an endotoxic liver failure model demonstrated a potentiation of LPS toxicity by GM-CSF. GM-CSF pretreated (10 min) mice died within 24 h of a challenge with a subtoxic dose of LPS, while control animals survived >72 h and monoclonal anti-GM-CSF antibody had a protective effect. GM-CSF potentiated systemic TNF release and hepatotoxicity induced by a subtoxic dose of LPS in galactosamine-sensitized mice. Polyclonal anti-GM-CSF IgG protected against septic liver failure and attenuated serum TNF concentrations. In vitro and ex vivo experiments revealed that LPS-induced IL-1 release from bone marrow or spleen cells was also enhanced (Tiegs et al. 1994). Therefore, GM-CSF seems to be an endogenous enhancer of LPS-induced organ injury, possibly by potentiating the release of pro-inflammatory cytokines such as TNF and IL-1.

When a high dose of GM-CSF was administered to sarcoma patients with neutropenia for a duration of two weeks, no increase in basal release of TNF-α or IL-1β by monocytes ex vivo was found, though the LPS-stimulated release of both factors reached 8-fold and 10-fold the respective values of day 0 (Perkins et al. 1993). GM-CSF-treated patients' monocytes produced more TNF than control patients' cells in response to tumor cells of the line U937 (Wiltschke et al. 1995). Thomassen et al. demonstrated that GM-CSF therapy of patients with lung cancer caused their monocytes and alveolar macrophages to increase their synthesis of TNF, IL-1 and IL-6 mRNA (Thomassen et al. 1991). They found differential response in cytokine secretion: the monocytes showed enhanced secretion of all three cytokines on the third day of treatment by continuous infusion, but the alveolar macrophages showed only enhanced IL-6 secretion by day 10 (Thomassen et al. 1991). Administration of GM-CSF augmented the release of TNF-α by alveolar macrophages ex vivo without altering their capacity to release reactive nitrogen intermediates (Mandujano et al. 1995). Furthermore the ex vivo cytokine secretion, i.e. TNF-α, IL-1β, IL-6 and IL-8, of peripheral blood monocytes was impaired, but their oxygen radical release was increased (Maurer et al. 1993).

Shortly after in vivo treatment of patients with GM-CSF, an increase in neutrophil and alveolar macrophage levels took place parallel to a rise in IL-6 and IL-8 levels in the bronchoalveolar lavage fluid (Gatti et al. 1995). In vivo treatment of volunteers with GM-CSF resulted in IL-8 release but no increased IL-6 or TNF-α serum levels (van Pelt et al. 1996). A single dose of GM-CSF resulted in the significant increase of in vivo plasma levels of sCD25 and IL-1ra as well as a trend to increased IL-8 levels (Aman et al. 1996). Patients with myelodysplasia treated in vivo with GM-CSF excreted more urinary cysteinyl leukotrienes than did untreated patients, indicating increased lipid mediator synthesis (Di Persio and Abboud 1992). A correlation to the expansion of the WBC was found. To investigate whether GM-CSF causes an activation of cellular immunity, urinary neopterin levels of gynecological cancer patients were assessed and found to have increased in response to GM-CSF (Marth et al. 1994). Neopterin, whose physiological role is unknown, is a sensitive marker of macrophages activated by interferons (Marth et al. 1994).

Taken together, although it induces the expression of the mRNA of various mediators, GM-CSF alone appears to induce only little monokine release. However, monocytes/macrophages become primed to respond to a variety of stimuli. As GM-CSF is known to be produced by monocytes spontaneously and on stimulation, GM-CSF thus might also have an autocrine function.

G
Expression of Surface Molecules

It was found that the expression of CD14, a known LPS receptor, on monocytes is down-modulated by GM-CSF in vitro (Kruger et al. 1996a). As with neutrophils, exposure to GM-CSF caused the rapid and complete loss of CD62L from the cell surface of monocytes in culture (Griffin et al. 1990). The surface expression of CD54 and CD18 was augmented in human blood monocytes and tumor-associated macrophages after incubation with GM-CSF, but CD50 (ICAM-3) was not influenced (Bernasconi et al. 1995). The IL-2 receptor was downregulated on human peripheral mononuclear cells in vitro by GM-CSF through the induction of PGE_2 (Hancock et al. 1988).

GM-CSF was shown to cause an increase in monocyte expression of surface HLA-DR and HLA-DP molecules important for monocyte T-cell interactions in vitro (Smith et al. 1990b; Chantry et al. 1990; Gerrard et al. 1990). In the presence of dexamethasone, GM-CSF caused a marked augmentation in HLA-DR, DP and DQ mRNA levels as well as surface expression on monocytes (Sadeghi et al. 1992). Overnight incubation of adherent murine spleen cells with antigen and GM-CSF led to more efficient antigen-presenting cells than without GM-CSF (Morrissey et al. 1987). The mechanism was found to lie in a GM-CSF-dependent increase in IL-1 secretion and Ia antigen expression (Morrissey et al. 1987). The antigen- presentation function of bone marrow-derived macrophages cultured with M-CSF was greatly enhanced after pulse treatment with GM-CSF (Fischer et al. 1988). GM-CSF enhanced macrophage accessory function in ConA-, mitogen- or antigen-stimulated T-cell proliferation in a dose dependent manner in vitro (Kato et al. 1990; Smith et al. 1990b). Incubation with GM-CSF also increased the number of CD32 receptors on monocytes (Liesveld et al. 1988), but antagonized the TGF-β-induced expression of CD 16 (Kruger et al. 1996b). Exposure of human monocytes to GM-CSF also induced the expression of CD23 (FcεRII) (Hashimoto et al. 1997). This effect could be abrogated by TNF. High, but not low, doses of GM-CSF down-modulate the expression and transcription of the M-CSF receptor on immature progenitors and monocytes (Panterne et al. 1996). This action was found to work via post-transcriptional mechanisms (Gliniak and Rohrschneider 1990). GM-CSF induces high levels of Ia expression in murine bone marrow-derived macrophages (Willman et al. 1989). Low, but not high, concentrations of GM-CSF result in a marked increase in the numbers of GM-CSF receptors on murine peritoneal exsudate macrophages (Fan et al. 1992). This effect was abrogated by the protein synthesis inhibitor cycloheximide.

The monocytes of healthy volunteers displayed increased expression of CD11b in response to a single dose of GM-CSF (van Pelt et al. 1996). Monocytes derived from patients receiving GM-CSF therapy show an significant increase in MHC class I and II antigen expression (Wiltschke et al. 1995; Aman et al. 1996). When cancer patients were given GM-CSF before and after chemotherapy, their peripheral blood monocytes displayed markedly stimulated expression of CD44 (Aman et al. 1996)

H
Microbial Killing

GM-CSF in synergy with TNF-α activated macrophages to kill or inhibit intracellular growth of *Mycobacterium avium*, when given both before and after establishment of infection (Bermudez and Young 1990). GM-CSF plays an important role as an endogenous mediator for the activation of macrophages by vitamin D_3 to kill or inhibit intracellular growth of these bacteria. Enhanced killing of avirulent *M. avium* by GM-CSF was demonstrated to be dependent on the generation of reactive nitrogen intermediates (Freund and Kleine 1992).

GM-CSF was also found to activate the intracellular killing of *Leishmania donovani* (intracellular parasitic protozoa) by human monocyte-derived macrophages. This antileishmanial effect did not depend on LPS as a secondary signal and reached maximal activation sooner than with IFN-γ (Weiser et al. 1987). Another group, using *Leishmania mexicana amazonensis*-infected macrophages, found a significant dose- dependent reduction of the parasites on incubation with GM-CSF and an even greater effect when a combination of GM-CSF with IFN-γ was used (Ho et al. 1990). GM-CSF further enhanced the growth inhibition of *Candida albicans* by fresh and aged human peripheral blood monocytes (Wang et al. 1989; Liehl et al. 1994).

I
Other

GM-CSF reduced the activity of the HIV enzyme reverse transcriptase in persistently infected monocytic U-937 cells. When GM-CSF was added to the medium before and during infection, it reduced reverse transcriptase activity by 90–100% and eliminated most viral antigen expression, but did not prevent return of productive infection after removal (Hammer et al. 1986; Freund and Kleine 1992).

3.2
Effects on the Functions of Lymphocytes

LAK cell function was generated in PBMC from cancer patients in vitro by brief exposure of the cells to IL-2 or, to a greater extent, through the synergy between low-dose IL-2 and GM-CSF (Baxevanis et al. 1995). RmuGM-CSF plays a role in the differentiation of murine B lymphocytes to fully functional antibody-secreting cells. In conjunction with IL-2, a plaque-forming cell (PFC) response by murine spleen cells was recorded (Grabstein et al. 1986b). A further study found that only splenic adherent cells and neither resting nor activated T or B cells expressed specific GM-CSF receptors, so splenic adherent cells were pulsed with GM-CSF before addition to macrophage-depleted cultures, where they reconstituted the PFC response to a greater degree than did control macrophages (Morrissey et al. 1987). In addition, GM-CSF augmented IL-2 production by splenic T-cells.

The GM-CSF-dependent growth of AML blast cells was enhanced by TNF; TNF and GM-CSF together induced AML blasts to produce IL-1β. GM-CSF increased the synthesis of IL-1ra by PMA- differentiated AML-193 cells (Kindler et al. 1990). In another study on AML blasts, IL-1 was found to stimulate the proliferation of blasts grown with GM-CSF in five of eight examined cases. This effect was inhibited by anti-TNF antibodies in four of the five cases (Ross and Koeffler 1992). These interactions were suggested to be useful for the treatment of leukemia.

4
Effects on the Functions of Other Cells

The main actions of GM-CSF are directed at leukocytes as displayed in Fig. 3. However, GM-CSF was also found to have effects on other cells. GM-CSF was found to specifically bind to normal human endothelial cells, activate the Na^+/H^+ antiport, and promote their proliferation (Di Persio 1990). GM-

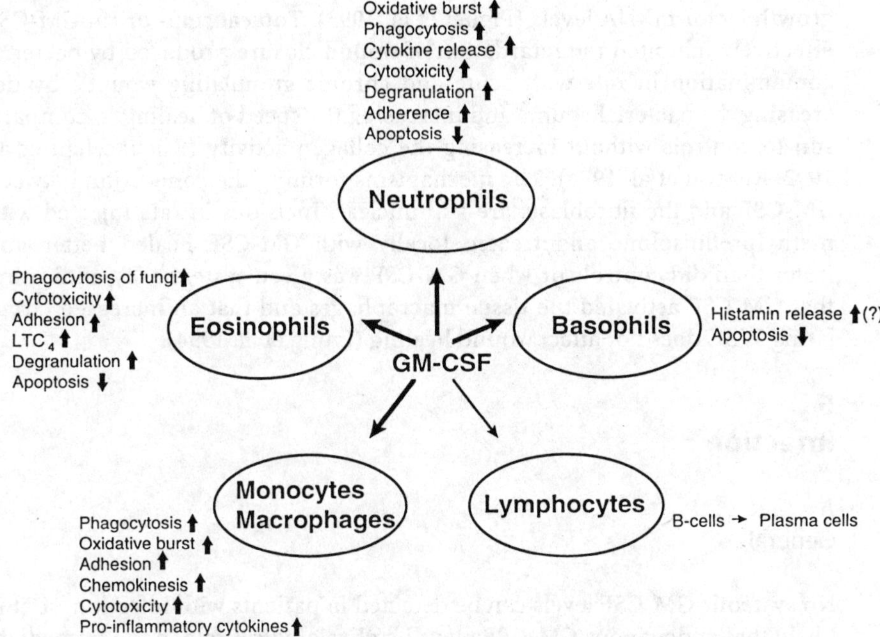

Fig. 3. Pertinent immunomodulatory effects of GM-CSF on different leukocyte populations. The diagramm summarizes the predominant effects of GM-CSF on immune functions of different types of leukocytes

CSF markedly enhanced the proliferation of liver endothelial cells of rats pretreated in vivo with LPS (Feder and Laskin 1994). It also plays a role in the proliferation and the migration of endothelial cells across gelatin-coated polycarbonate filters (Bussolino et al. 1989).

The proliferative effect of GM-CSF also extends to tumor cell lines: small-cell lung carcinoma cell lines, which have high-affinity GM-CSF receptors, were shown to be responsive, as were two osteogenic sarcoma cell lines, a breast carcinoma cell line, a SV40-transformed marrow stromal cell line and colon adenocarcinoma cell lines (Gasson 1991). To date, however, there is no evidence of tumor promotion in cancer patients treated with GM-CSF.

Animal data also suggest a role for GM-CSF in wound healing: In mice instilled with bleomycin to elicit pulmonary fibrosis, a significant increase in GM-CSF mRNA was recorded on day 5 after bleomycin administration. GM-CSF had an inhibitory influence on the alveolar remodeling and collagen deposition associated with pulmonary fibrosis. Anti-GM-CSF antibody had opposite effects to GM-CSF and decreased TNF-α mRNA levels as seen on day 15, but otherwise had no influence on TGF-β, IL-1α or platelet derived growth factor mRNA levels (Piguet et al. 1993). Topical rmu- or rhuGM-CSF effectively inhibited the retardation of wound closure produced by bacterial contamination in rats with acute and chronic granulating wounds by decreasing the bacterial counts and increasing the speed of healing in comparison to controls without increasing the collagen activity (Kucukcelebi et al. 1992; Robson et al. 1994). The mechanisms forming the connection between GM-CSF and the fibroblasts are still unclear. Incisions in rats injected with methylprednisolone and treated locally with GM-CSF healed better and faster than did controls or when GM-CSF was given systemically, indicating that GM-CSF activated the tissue macrophages and that an increased circulating WBC does not affect wound healing (Jyung et al. 1994).

5
Infection

A
General

No systemic GM-CSF levels can be detected in patients with infection, so it is likely that endogenous GM-CSF plays its physiological role in the immediate vicinity of the cells by which it is secreted (Freund and Kleine 1992). The exogenous administration of GM-CSF is aimed at stimulating nonspecific resistance to potentially lethal infections by increasing the WBC. But to fulfill these expectations, the induced monocytes and neutrophils must be functionally mature and primed for microbial killing.

The infection of GM-CSF-pretreated, myelosuppressed mice with normally lethal doses of *Pseudomonas aeruginosa*, *Staphylococcus aureus* or *Candida albicans* resulted in a significant dose-dependent improvement of survival (Tanaka et al. 1989; Liehl et al. 1994). There has also been research on combinations of GM-CSF with IL-6 or leukemia inhibitory factor (LIF), which are both potent inducers of the acute-phase response and can induce an increase in platelet counts (Liehl et al. 1994). The rationale for these combinations is that the synergism of the induction of opsonization of microorganisms by acute phase proteins with the activation of phagocytes by GM-CSF should increase resistance to infections. This hypothesis was proven to be correct when myelosuppressed mice were treated with either of the combinations and infected with *Pseudomonas aeruginosa* (Liehl et al. 1994).

Prophylactic rmuGM-CSF given i.p. six hours before a 90% lethal dose challenge of *S. aureus* significantly improved the survival in a neonatal rat model of infection (Frenck et al. 1990). Neonatal rats, which have been found to have deficient PMN production and function in infection, with streptococcal sepsis were given rhuGM-CSF after infection. A higher survival rate than control animals not given rhuGM-CSF was reported, apparently caused by phagocyte priming and/or cellular influx into the peritoneum (Wheeler and Givner 1992), although human GM-CSF is generally believed to be non cross-reactive in mice. GM-CSF administered in conjunction with penicillin to neonatal rats with established group B streptococcal infection decreased the mortality rate substantially in comparison to penicillin alone (Givner and Nagaraj 1993). The survival in two mouse models of gut-derived sepsis was improved by GM-CSF through better gut barrier function and better bacterial clearance (Gennari et al. 1994). However, we found that the prophylactic administration of rmuGM-CSF neither augmented leukocyte numbers nor protected mice from lethal fecal peritonitis (Barsig et al. 1996).

Rats with sepsis-induced organ injury by cecal ligation and puncture given rmuGM-CSF showed no increased survival rates, but rather earlier deaths than the control group; an inhibition of early leukosequestration in the peritoneal cavity was observed (Toda et al. 1994). In addition to the mild hepatic injuries observed in shock, the livers of GM-CSF-treated rats showed centrilobular degeneration and necrotic changes. These rats also had an earlier increased plasma alanine-α-ketoglutarate-aminotransferase (ALT) level, suggesting that liver dysfunction may have contributed to their earlier death (Toda et al. 1994). The inhibition of neutrophil migration by systemic administration of GM-CSF and its stimulating activity on the expression of TNF might play a role in the process. Therefore, the systemic application of GM-CSF, as opposed to its natural local production, might be dangerous when infection is already present.

In a murine model of 20% surface burn plus cecal ligation and puncture on day 10 after injury, survival was significantly better in animals treated with GM-CSF on days 5–9 after the burn (Molloy et al. 1995). ConA-stimulated T-cell proliferation and IL-2 production, which were suppressed after burn injury, were also improved by treatment with GM-CSF (Molloy et al. 1995).

Mycobacterium avium complex synthesizes superoxide dismutase, which can inactivate macrophage-derived superoxide anions as well as enzymes which can hinder their production (Bermudez 1994). However, the stimulation of *M. avium* infected murine and human macrophages with GM-CSF induced increased ROS production and enhanced mycobacteriostatic/mycobactericidal activity ex vivo (Denis 1991a; Bermudez 1994). GM-CSF and TNF-α had additive effects on bacteriostatic activity of macrophages towards *M. avium* (Denis 1991a).

B
Specific Pathogens

The in vitro findings that GM-CSF and TNF-α synergize to allow peritoneal macrophages to overcome the anti-phagocytic properties of virulent yeast were studied further by administering neutralizing monoclonal antibody specific for GM-CSF to *Cryptococcus neoformans*-infected mice. However, this increased mortality and induced the rapid progression of the disease (Collins et al. 1992).

Endogenous GM-CSF plays a role in the initial host defense response to *Leishmania donovani* as could be seen by the upregulation of GM-CSF mRNA in infected mouse livers and the exacerbation of infection when the anmals were treated with anti-GM-CSF antibody (Murray et al. 1995). In an animal model, GM-CSF was found to reduce the proportion of cultured peritoneal macrophages infected with *Leishmania tropicana* collected from mice (Scarffe 1991). There was an indication of increased killing of the parasites. But, GM-CSF may also play a role in the pathogenesis of *Leishmania* infection itself: murine lung-conditioned medium has colony-stimulating activity on bone marrow cells and on promastigotes of *Leishmania mexicana amazonensis*. Both activities are abrogated by immunoprecipitation with an anti-murine-GM-CSF antibody. The hypothesis that GM-CSF plays a part in the *Leishmania* pathogenesis is supported by the fact that T-lymphocytes and macrophages, both GM-CSF producers, actively participate in the leishmanial infection (Freund and Kleine 1992). *Leishmania donovani* infection of bone marrow-derived macrophages was found to induce gene expression of, among others, GM-CSF and TNF, which were shown to inhibit apoptosis of these macrophages induced by the removal of M-CSF (Moore et al. 1994).

Mice chronically infected with *Trypanosoma cruzi* (Chagas disease) have a defect in their cellular immune response, that was corrected by the administration of GM-CSF in vivo. The effect seemed to depend on an enhancement of the expression of Ia antigen and IL-2 mRNA in accessory cells of the spleen and on the stimulation of IL-1 production in peritoneal macrophages (Reed et al. 1990). GM-CSF can inhibit the replication of *Trypanosoma cruzi* by activating macrophages in both human and murine cultures and increasing their ability to release ROS (Reed et al. 1987).

Neurologic damage from cerebral malaria is mediated by high local levels of TNF. Anti-TNF antibodies are protective (Ross and Koeffler 1992), but there seems to be an interaction between GM-CSF and TNF in this disease, as antibodies against both GM-CSF and IL-3 are also protective in mice infected with *Plasmodium berghei* (Grau et al. 1988), by preventing the development of high TNF levels. However, antibodies against GM-CSF or IL-3 alone have no effect. It is thought that GM-CSF and IL-3 effect the production of macrophages and cause them to accumulate in infected tissue, while IFN-γ and *Plasmodium* products activate them to produce TNF.

Pneumocystis carinii pneumonia (PCP) is a major cause of morbidity and mortality in patients with AIDS. In mice depleted of $CD4^+$ T-lymphocytes and inoculated with *P. carinii,* the intensity scores of PCP infection were significantly decreased in mice treated with rmuGM-CSF in comparison to control mice, which may be due to an increased production of TNF-α by alveolar macrophages, which has been shown to bind and kill *P. carinii* in vitro (Mandujano et al. 1995). These results suggest that impaired cytokine production rather than an intrinsic macrophage defect may predispose AIDS patients to opportunistic infections. GM-CSF therapy could reverse the relative deficiency of endogenous GM-CSF in such patients and enhance the cytotoxic activity of macrophages and the release of cytokines important for host defense (Mandujano et al. 1995).

RmuGM-CSF has been found to protect mice against *Sendai* virus infection when given intranasally one or three days before the infection (Freund and Kleine 1992). GM-CSF also showed protective activity when administered s.c. to cyclophosphamide-treated mice before i.v. infection with herpes simplex virus (Freund and Kleine 1992).

In summary, GM-CSF administered prophylactically seems to be beneficial in animal infection models in most cases. But, when severe active infections are already present, it has detrimental effects when administered alone, though it may be a useful adjuvant to antibiotic therapy. In contrast to G-CSF, GM-CSF appears to act as a pro-inflammatory factor. A further indication for GM-CSF may be to increase eosinophil cell counts to combat parasitic infection. Here, GM-CSF has an advantage over G-CSF, as it also activates monocytes and macrophages. If enhanced macrophage activation including the production of IL-1 and TNF is desired, GM-CSF may again be indicated (Gabrilove and Jakubowski 1990). To date, too little is known about a possible benefit of GM-CSF in the treatment of viral infections.

6 Potential Clinical Applications

Nissen and Hovgaard formulated the clinical uses of the CSF such as GM-CSF: 1. to increase proliferation of normal bone marrow cells in primary marrow disorders, such as aplastic anaemia, cyclic neutropenia, myelodysplastic syndromes and in secondary-induced marrow hypofunction, e.g. after conventional chemotherapy, bone marrow transplantation and other drug-induced cytopenias; 2. to increase the number of circulating stem cells; 3. to recruit tumor cells into S-phase of cell cycle to increase their susceptibility to subsequent cytotoxic agents; 4. to increase effector cell function, e.g. anti-neoplastic effect, anti-infective effect and reversal of leukocyte function defects (Cairo et al. 1992a).

To date, the predominant clinical use for GM-CSF (similar as for G-CSF) is in neutropenia associated with cancer therapy. After the rapid recovery of neutrophil counts was demonstrated in myelosuppressed monkeys (Liehl et al. 1994), the efficacy of GM-CSF was evaluated in patients with cancer chemotherapy-induced myelosuppression (Antman et al. 1988), patients who had been accidentally exposed to cesium-137 (Ruef and Coleman 1990) and patients undergoing bone marrow transplantation (Liehl et al. 1994). Now, GM-CSF is used extensively to decrease the severity and duration of leukopenias, as reviewed by Moore (1991).

The effectiveness of GM-CSF does depend on a sufficiently large population of target cells, so it is not surprising that in progressed stages of aplastic anaemia with corresponding hypocellularity of the bone marrow, the increase in bone marrow cellularity and peripheral leukocyte counts were disappointing where earlier stages showed greater effects (Ruef and Coleman 1991). Combination of GM-CSF with factors acting earlier on in hematopoiesis such as IL-3 or IL-1 might be more beneficial (Moore 1991). But, as there is great interindividual variation in the response to the therapy, its use still seems justified in progressed cases (Ruef and Coleman 1991).

In patients with myelodysplastic syndromes, GM-CSF therapy increased the WBC, but some patients with a certain level of marrow blasts developed AML (Lieschke and Burgess 1992a). Here, a different approach is to administer GM-CSF concurrently with cytorabine in an attempt to increase the susceptibility of malignant cells to cytotoxic therapy (Lieschke and Burgess 1992a).

Furthermore, short treatment with GM-CSF before chemotherapy could reduce the hematopoietic toxicity of cytostatics and thereby also enabled the dose intensity of protocols to be increased (Aglietta et al. 1993). This was investigated on the basis of observations that the suspension of GM-CSF treatment resulted in a reduction of the proliferative activity of the hyperplastic marrow to values below the baseline after three days, suggesting refractoriness of hematopoietic progenitors to the action of cell cycle-specific cytostatic agents (Aglietta et al. 1993).

The possibility of augmenting GM-CSF's effect on the WBC still further by administering it in combination with another growth factor was tested on a normal rhesus monkey, that was given IL-3, known to induce enhanced cellularity and increase of progenitors in the bone marrow, and then GM-

CSF. The peripheral blood counts of all colony types evaluated were substantially increased over the levels observed with GM-CSF or IL-3 monotherapy (Liehl et al. 1994). In vivo, the combination IL-3/GM-CSF resulted in a high number of biologically active progenitor cells as shown by a rapid engraftment of neutrophils and platelets after autologous transplantation (Kanz 1994).

The rationale behind the use of GM-CSF in AML, beside stimulating regenerating hematopoiesis, is to improve the efficacy of chemotherapy by triggering leukemic progenitor cells into cycle (Hast et al. 1995). Leukemic progenitor cells express GM-CSF receptors on their surface and can be recruited into cycle by cytokine stimulation both in vitro and in vivo. This can be exploited by employing the S-phase specific cytarabine. In vitro, GM-CSF potentiated the cytotoxicity of cytarabine by increasing the number of cells in the S-phase (Scarffe 1991). The metabolism of cytarabine, which is necessary for its full cytotoxic activity, seemed to be antagonized by GM-CSF to a greater degree in normal cells than in leukemic cells, which should result in improved selectivity of cytarabine for leukemia.

GM-CSF therapy was further found to increase the number of complete remissions in patients with non-Hodgkin's lymphoma significantly (Gerhartz et al. 1995). As the improved adherence to the chemotherapy schedule was not thought to be sufficient to evoke such high response rates, it was postulated that direct effects of GM-CSF might also play an important role in tumor response by increasing the susceptibility of some cell populations to cytolysis by LAK cells; by enhancing effector cell phagocytosis, possibly mediated by increased superoxide generation and increased cytokine production by such cells; or by enhancing ADCC (Gerhartz et al. 1995).

Beside reversal of neutropenia by administration of CSF, these factors also have a potential role in bone marrow transplantation: G-CSF and GM-CSF not only support engraftment but are also increasingly used for collection of stem cells from peripheral blood. The ability of GM-CSF to induce the substantial mobilization of progenitor cells into the peripheral blood (Liehl et al. 1994) allowed the induction of sufficient peripheral blood stem or progenitor cells to harvest them for autotransplantation (Scarffe 1991; Liehl et al. 1994; Kanz 1994). This procedure enabled a further escalation of the intensity of the chemotherapy where a relationship between this and tumor response had been defined. Positive selection of $CD34^+$ peripheral blood progenitor cells reduced the number of potentially contaminating tumor cells within the preparations of patients with solid tumors (Kanz 1994). Patients receiving progenitors collected with GM-CSF demonstrated faster recoveries of total and neutrophil counts compared with the control group and a recovery in platelet counts and a reduction in the severity of mucositis have been reported (Scarffe 1991). Further studies suggest that the engraftment of peripherally collected stem cells, augmented by GM-CSF may be so reliable that concomitant bone marrow transplantation is no longer required (Scarffe 1991). The use of GM-CSF in the collection of peripheral blood stem cells promises to reduce the requirement for extensive support measures such as sterile environments, antibiotic therapy and platelet transfusions, making intensive chemotherapy cheaper and safer (Scarffe 1991).

In conclusion, the clinical use of GM-CSF in neutropenic patients is very similar to the one of G-CSF. However, the fact that only one tenth of patients receives GM-CSF while G-CSF is more broadly used is only partially a result of marketing strategies. Due to the pro-inflammatory nature of GM-CSF in contrast to the anti-inflammatory characteristics of G-CSF, GM-CSF therapy was found to be associated with many more severe side-effects. Therefore, additional indications have been targeted where the immunostimulatory efficacy of GM-CSF beyond mere leukocyte recruitment is favourable.

One of these leukocyte-stimulating activities might be tumor cell cytotoxicity of leukocytes, which is partially induced by the pro-inflammatory mediators increased by GM-CSF. The potential of GM-CSF in activating macrophages and monocytes to destroy cytokeratin (CK2) positive cells (micrometastases) in the bone marrow of patients with gastric cancer when tested in a very small study was promising: signs of monocyte activation and a decrease in the number of CK-positive cells in the bone marrow were observed (Schlimok et al. 1995). There was also a positive trend in the relapse rate for patients treated with GM-CSF.

The rationale for the use of GM-CSF in leukemic patients receiving T-cell-depleted allogeneic bone marrow transplantation is based on properties of GM-CSF of inducing potential anti-tumor mechanisms in monocytes, such as the secretion of pro-inflammatory factors and the stimulation of ADCC. If GM-CSF exerts these anti-leukemic effects, it should accelerate engraftment and minimize the incidence of leukemic relapse after removal of T-cells (De Witte et al. 1995). There are also studies suggesting that GM-CSF will improve the outcome for patients with graft failure without exacerbating graft-versus-host disease (GVHD) (Scarffe 1991). A small study suggested that GVHD might actually be ameliorated by the use of GM-CSF (Lang et al. 1987).

Monocytes matured to macrophages with GM-CSF display similar receptors and tumoricidal activity to those differentiated in the presence of autologous lymphocytes (Lopez et al. 1993). When activated with IFN-γ, these cells can be used for adoptive immunotherapy trials (Lopez et al. 1993).

Another, positive aspect of the immunostimulatory activity of GM-CSF could be improved host defense, which might have a bearing for the treatment of infectious diseases. Various clinical studies have shown that therapy with GM-CSF significantly reduced the incidence of infections from either bacterial or fungal sources due to neutropenia, thereby reducing the demand for antibiotics and allowing the therapy to proceed according to schedule, e.g. with non-Hodgkin's lymphoma (Scarffe 1991; Gerhartz et al. 1994; Gerhartz 1995) or in the case of bone marrow transplantation (Moore 1991; Gulati et al. 1995; De Witte et al. 1995; Gorin et al. 1995) and heightened the response rates to antibacterial and perhaps even antifungal treatment (Bodey et al. 1994). A case has been reported where a patient in intensive care with acquired agranulocytosis and sepsis experienced rapid neutrophil recovery and resolution of a clinical infection when treated with GM-CSF (Weiss et al. 1992).

Yet, in a small study of patients receiving intensive chemotherapy for lymphoma, the number of infections acquired by patients receiving GM-CSF only was greater than that of patients receiving prophylactic antibiotics (Scarffe 1991), suggesting that GM-CSF can be used as an adjunct to antibiotics, but not instead of them.

The migration of allogeneic transfused neutrophils to sites of a *Candida albicans* infection was reported to be effective in a patient receiving GM-CSF and pentoxifylline (Montgomery et al. 1991). Pentoxifylline decreased TNF-α levels in vivo (Montgomery et al. 1991) and prevented and reversed the ability of TNF-α to inhibit neutrophil migration in vitro while decreasing CD11b expression (Montgomery et al. 1991). In addition, pentoxifylline inhibited GM-CSF's induction of TNF-α in early myeloid cells without affecting proliferation (Montgomery et al. 1991), which makes the combination of these two agents in clinical practice attractive.

Acute visceral leishmaniasis causes a suppression of the immune system. The patients demonstrate an abnormal production of Th_1 lymphocyte-derived macrophage-stimulating factors, such as IL-2, IFN-γ, GM-CSF and IL-3, that may be explained with the increased production of Th_2 cytokines, such as IL-4 and IL-10, which downregulate the anti-leishmanial response (Badaró et al. 1994). A patient treated with GM-CSF plus antimony produced IFN-γ, which two patients receiving only antimony did not do, suggesting that GM-CSF may function to alter the immune balance in patients with acute visceral leishmaniasis. This hypothesis must still be verified.

GM-CSF was effective for the stimulation of nonspecific protection against *Sendai* virus or herpes simplex type 1 virus (HSV) in mice (Azuma et al. 1992).

A variety of clinical conditions are associated with functional impairment of leukocytes. CSF also have the potential to reverse this functional leukopenia, e.g. in HIV infection, in newborn and preterm infants or following trauma, burn or sepsis. Qualitative and quantitative defects in neutrophils and monocytes in AIDS patients, which may be exacerbated by antiviral therapy with zidovudine probably contribute to the high incidence of opportunistic infections and neoplasms (Scarffe 1991). Therapy with GM-CSF increased the numbers of granulocytes, monocytes and eosinophils, enhanced the anti-HIV-activity of zidovudine as well as the ADCC and phagocytic capacity of patient neutrophils; superoxide generation capacity was enhanced in response to a chemotactic agent (Moore 1991; Mitsuyasu 1994). Although the functions of neutrophils from most patients were found to be normal and normally enhanced by GM-CSF, abnormalities of neutrophils from patients with phagocytosis or intracellular killing were corrected by GM-CSF infusion in vivo (Baldwin et al. 1988; Scarffe 1991; Mitsuyasu 1994), suggesting that GM-CSF might improve control or prophylaxis of opportunistic infections and possibly mortality and survival. Furthermore, neutrophils produced in vivo in response to GM-CSF functioned normally and showed signs of priming and activation in vivo (Baldwin et al. 1988). Nevertheless, a number of opportunistic infections developed in patients receiving GM-CSF (Moore 1991). GM-CSF plus ganciclovir given to patients with AIDS-associated cytomegalovirus retinitis and bone marrow intolerance to the antiviral agent resulted in normal ANC with no resistance to ganciclovir and prevention of retinits progression in virtually all surviving, evaluable patients (Scarffe 1991). Whether the benefits GM-CSF, as shown in animal models of opportunistic infections common in the HIV-infected population can also be reaped in a clinical setting with HIV-infected patients, remains to be seen (Frumkin 1997). When GM-CSF and zidovudine therapy were alternated, an increase in HIV replication was observed, but in combination therapy this effect was absent and there was a decrease in HIV (p24)-antigen in the blood (Ruef and Coleman 1991). In vitro studies suggest that GM-CSF stimulates the replication of HIV in monocytes, but that the antiviral effect of zidovudine is enhanced by GM-CSF (Lieschke and Burgess 1992a). Although ADCC can be mediated by various classes of lymphocytes and monocytes, these may also be infected with HIV, therefore it seems better to focus on the neutrophils, as these are apparently not productively infected by the virus, i.e. G-CSF seems to be the therapeutic cytokine of choice (Baldwin et al. 1989; Hengge et al. 1992).

Severe trauma and sepsis results in states of suppressed immune functions. First in vitro studies are promising that GM-CSF might be suitable to treat this type of immunodeficiency (Randow et al. 1997). When treating these patients, additional effects beyond reactivation of paralyzed immune cells by recruitment of new leukocytes from bone marrow can be expected. Burn patients treated with GM-CSF displayed an increased WBC, normal stimulated oxidative function, increased MPO activity which returned to normal after a week and normal oxidative burst activity compared with patients not receiving GM-CSF (S116).

The immunostimulatory effect of GM-CSF might be useful as a vaccine adjuvant: When monkeys were treated with rhGM-CSF and IL-3, their antibody titers to IL-3 were substantially greater than with IL-3 monotherapy (Liehl et al. 1994). Mice immunized with bovine serum albumin (BSA) and given GM-CSF developed a substantially increased antibody response compared with mice not given GM-CSF (Liehl et al. 1994). The adjuvant effect of GM-CSF might, however, also be a source of side-effects: In a study of ovarian cancer patients receiving GM-CSF, two patients who had demonstrated a low titre of anti-thyroid antibodies before the study showed an increase in antibody titre and transient thyroiditis after administration of GM-CSF, which suggests a systemic effect of GM-CSF on primary and secondary antibody responses. When antibody corresponding to a specific idiotype expressed on B cell lymphomas was fused to GM-CSF and injected into mice with B-cell lymphoma xenografts, the mice developed antibodies to the lymphoma which showed a protective effect towards disease progression. These antibodies were not produced when the two molecules were injected separately at the same site. Preliminary clinical trials suggest that GM-CSF enhances antibody responses to hepatitis B vaccine (Jones et al. 1994).

Taken together, a number of putative indications for GM-CSF beyond reversal of neutropenia are emerging. These new opportunities are closely linked to the immunostimulaory capabilities of GM-CSF. In each case, however, the benefit of immunostimulation and the risk of adverse effects must be weighed up. For example, the fact that GM-CSF primes bone marrow precursors for enhanced leukotriene synthesis (Di Persio and Abboud 1992) and macrophage TNF-α release (Thomassen et al. 1991, Wiltschke et al. 1995) may be associated with the greater systemic toxicities incurred by patients when given GM-CSF in vivo. Increased adhesion and increased toxic radical production by granulocytes have been strongly implicated in the pathogenesis of ARDS, yet GM-CSF was not implicated in the pathogenesis of neonatal RDS, as the GM-CSF plasma levels of infants who developed the syndrome were similar to a control group; both were higher than adult levels (Yurdakök et al. 1994). Nevertheless, GM-CSF will have to be explored much more carefully in patients prone to sepsis than the anti-inflammatory G-CSF.

III
M-CSF

1
General Information

1.1
Molecular Biology and Endogenous Production

A
The Molecule M-CSF

The molecular biology of M-CSF is the most complex of all CSF. The M-CSF gene (22 kb), which is located on the short arm of chromosome 5 in humans, has 10 exons and 9 introns. Alternate splicing within exon 6 produces 3 forms of M-CSF mRNA that are translated into M-CSF-α, -β and -γ. The signal sequence (32a.a.), the biologically active part of the molecule (149a.a.), a transmembrane domain (23a.a.) and a cytoplasmic domain (36a.a.) are common. Genomic and cDNA clones reveal that the difference between the forms is the length of an insertion sequence after amino acid 181. The N-terminal domain has two N-linked glycosylation sites and seven cysteine residues. The M-CSF-α soluble protein is a disulfide-bonded homodimer with a molecular weight of 47–56 kd, whereas M-CSF-β and -γ secreted proteins are disulfide-bonded homodimers of 88 kd (Herrmann et al. 1990). Removal of the transmembrane and cytoplasmic domains by deletion mutagenesis abolished M-CSF surface expression and favored a more efficient secretion of the protein, suggesting that M-CSF is normally synthesized as a membrane-bound precursor that is proteolytically cleaved to release the soluble form (Heard et al. 1987).

Human M-CSF binds to the murine M-CSF receptor but not vice versa (Hamblin 1988).

B
Endogenous Production of M-CSF

M-CSF gene expression is primarily controlled posttranscriptionally (Ernst et al. 1989; Herrmann et al. 1990). In vitro, activated monocytes (Rambaldi et al. 1987), neutrophils (Lindemann et al. 1989b), bone marrow stromal cells (Gualtieri et al. 1987; Fibbe et al. 1988), epithelial cells of the endometrium of pregnant uterus (Pollard et al. 1987) and of the thymus (Galy et al. 1993), endothelial cells (Cannistra and Griffin 1988), mesothelial cells (Lanfrancone et al. 1992) and smooth muscle cells (Filonzi et al. 1993) produce M-CSF. Human neuroblastoma cells and murine neurons, namely granule cells of the cerebellum, also might produce M-CSF; the mRNA, but not the protein, is present in mouse brain including cerebellum (Nohava et al. 1992).

Expression of M-CSF is also subjected to autocrine regulation (Herrmann et al. 1990): Monocytes, which are a major source of M-CSF, also bear high-affinity M-CSF receptors. Both transcripts of M-CSF and its receptor transcripts are induced during monocytic differentiation of human HL-60 leukemia cells, suggesting that they can regulate their own survival, growth and differentiation (Horiguchi et al. 1986).

Induction of M-CSF

The levels of M-CSF mRNA in resting monocytes were found to be very low (Horiguchi et al. 1986). Stable M-CSF mRNA was only produced by monocytes in vitro in response to an exogenous signal (Rambaldi et al. 1987; Oster et al. 1989c). Treatment of monocytes with PMA, cycloheximide, IL-3, GM-CSF (Rambaldi et al. 1987; Ernst et al. 1989), PMA plus PHA (Oster et al. 1989c), IFN-γ (Rambaldi et al. 1987; Gruber and Gerrard 1992), phorbol ester (TPA) (Horiguchi et al. 1986), TNF-α (Gruber and Gerrard 1992) or IL-4 (Wieser et al. 1989) in vitro induced the accumulation of M-CSF transcripts in the cells; the induction of the gene was due to mRNA stabilization, not increased transcription. The release of the biologically active protein was induced by PMA (Rambaldi et al. 1987; Gruber and Gerrard 1992), IL-3 (Vellenga et al. 1988) or GM-CSF (Horiguchi et al. 1987; Vellenga et al. 1988; Gruber and Gerrard 1992; Hamilton 1994) in monocytes. Stimulation of human monocytes with immobilized monoclonal antibodies directed against the CD45, CD44 or LFA-3 antigen induced the stabilization of mRNA transcripts and the production of M-CSF protein (Gruber et al. 1992). M-CSF mRNA and protein secretion became detectable when monocytes were cultured in the presence of TNF-α, but under identical conditions TNF-β failed to induce M-CSF synthesis (Oster et al. 1987). The enhanced release of M-CSF in response to TNF-α by human monocytes was confined to the leu $M3^+$, $HLA\text{-}DR^+$ population of cells (Lu et al. 1988a). LPS could enhance M-CSF formation in adherent or elutriated monocytes (Lee et al. 1990; Hamilton 1994), but the levels were influenced by cyclooxygenase products. In contrast, LPS-induced monocytes in nonadherent cultures did not express M-CSF (Lee et al. 1990).

PMN accumulated M-CSF mRNA in response to GM-CSF and were shown to translate this mRNA into secretory protein (Lindemann et al. 1989b). Both IL-1 and TNF-α induced an accumulation of M-CSF mRNA in endothelial cells in vitro; simultaneous treatment with both factors had an additive effect, suggesting that they acted via different pathways (Seelentag et al. 1987). Untreated internal mammary artery and aortic smooth muscle cells make M-CSF constitutively (Filonzi et al. 1993). IL-1, TNF-α and, in addition, IFN-γ raised the M-CSF levels produced by these cells. Bone marrow stromal cells produce M-CSF constitutively, but this action could be supported by the addition of IL-1 in vitro (Fibbe et al. 1988). Human peritoneal mesothelial cells constitutively produce M-CSF (Lanfrancone et al. 1992), an effect which was not altered by the addition of IL-1. Thymic epithelial cells produce M-CSF constitutively (Galy et al. 1993). Their production was up-regulated by IL-1 or to a lesser extent by IFN-γ; IL-4 had no effect (Galy et al. 1993). The expression of an alternatively spliced M-CSF mRNA by murine uterine glandular epithelial cells was regulated by the synergistic action of female sex steroids, estradiol-17β and progesterone (Pollard et al. 1987). See Table 8 for a summary of the factors influencing the production of M-CSF.

Modulation of M-CSF production

GM-CSF induced the production of M-CSF by monocytes, an effect that was enhanced by IFN-γ, M-CSF or TNF-α, of which TNF-α synergized with GM-CSF in this respect (Gruber and Gerrard 1992). M-CSF release by monocytes in response to TNF-α noted in another study was synergistically enhanced by IFN-γ. Further experiments showed that pre-incubation of the monocytes with TNF-β before TNF-α abolished the inductive effects of TNF-α, suggesting that it plays a role as an antagonist to TNF-α (Oster et al. 1987). The stimulatory action of the antibodies against CD45, CD44 or LFA-3 was dramatically augmented by LPS, IL-1β, but not IL-6 or TNF-α, though neither of these was able to induce M-CSF secretion alone in the absence of antibody (Gruber et al. 1992; Gruber and Gerrard 1992). M-CSF production by purified human monocytes in the presence of LPS was upregulated by cyclooxygenase inhibition by indomethacin; this effect is canceled by PGE_2 (Lee et al. 1990; Hamilton 1994). In LPS- treated cells, IL-4 as well as dexamethasone lowered the amount of M-CSF produced by purified human monocytes (Hamilton 1994). Monocytes and macrophages infected with HIV downregulated M-CSF production, an effect which may provide part of the explanation of the immunological dysfunction in HIV-infected patients (Esser et al. 1996).

Cycloheximide potentiated the effects of IL-1 and TNF-α on M-CSF mRNA levels of arterial smooth muscle cells. These results suggest that cytokine-stimulated arterial smooth muscle cells may be a source of the M-CSF found in arteriosclerotic lesions (Filonzi et al. 1993).

Table 8. Factors affecting the production of the M-CSF protein

Stimulus type	Stimulus	Cell type	Results	References
Constitutive	none	monocytes	–	Horiguchi et al. 1986
		stromal cells	+	Fibbe et al. 1988
		mesothelium	+	Lanfrancone et al. 1992
		smooth muscle cells	+	Filonzi et al. 1993
		epithelium	+	Galy et al. 1993
Bacterial molecules	LPS	monocytes	+	Lee et al. 1990
			–	Gruber et al. 1992
Pro-inflammatory mediators	IL-1	monocytes	–	Gruber et al. 1992
		stromal cells	+	Fibbe et al. 1988
		endothelium	+	Seelentag et al. 1987
		smooth muscle cells	+	Filonzi et al. 1993
		epithelium	+	Galy et al. 1993
	TNF-α	monocytes	+	Oster et al. 1987
			–	Gruber et al. 1992
		endothelium	+	Seelentag et al. 1987
		smooth muscle cells	+	Filonzi et al. 1993
	IFN-γ	smooth muscle cells	+	Filonzi et al. 1993
		epithelium	+	Galy et al. 1993
Colony stimulating factors	GM-CSF	monocytes	+	Horiguchi et al. 1987
		PMN	+	Lindemann et al. 1989b
	IL-3	monocytes	+	Vellenga et al. 1988
Other	IL-4	epithelium	–	Galy et al. 1993
	IL-6	monocytes	–	Gruber et al. 1992
	PMA	monocytes	+	Rambaldi et al. 1987
	PMA+PHA	PBMC	+	Oster et al. 1989c
	TNF-β	monocytes	–	Oster et al. 1987
	anti-CD44	monocytes	+	Gruber et al. 1992
	anti-CD54	monocytes	+	Gruber et al. 1992
	anti-LFA-3	monocytes	+	Gruber et al. 1992
	oestradiol 17-β	epithelium	+	Pollard et al. 1987
	progesterone	epithelium	+	Pollard et al. 1987

Key: + : induction of M-CSF production and secretion
– : no effect on the production of M-CSF

The independent regulation of M-CSF by different humoral signals may be physiologically relevant: chronic fungal infections that are associated with enhanced T-cell-mediated immunity result in the secretion of GM-CSF and IL-3 from activated T-cells which selectively induce M-CSF secretion by monocytes which in turn stimulates the monocytosis often observed during fungal disease or tuberculosis (Cannistra and Griffin 1988). See Table 9 for a list of the modulators of M-CSF production.

Serum M-CSF

After mice were infected with *Listeria monocytogenes*, the bulk of elevated serum colony-stimulating activity represented M-CSF and G-CSF, the degree of elevation depended on the infecting inoculum and the numbers of bacteria growing in the spleens and livers of the two mouse strains compared (Cheers et al. 1988).

Radioimmunoassay (RIA) results of 10 normal individuals suggest that endogenous circulating M-CSF is present at a low but detectable concentration (Shadle et al. 1989). M-CSF is elevated in febrile neutropenic or non-neutropenic patients in comparison to afebrile subjects and remains so for up to 10 days after resolution of infection (Cebon et al. 1994). M-CSF elevation in sepsis is correlated with fever, renal impairment and known pathogen (Cebon et al. 1994).

Deficient or excess M-CSF

Mice with an inactivating mutation of the M-CSF gene were toothless because of failure of incisor eruption, had skeletal abnormalities, developed osteopetrosis (op/op), were low in body weight and had compromised fertility (Wiktor-Jedrzejczak et al. 1991; Lieschke et al. 1994b). They displayed a major deficiency in macrophage-derived osteoclasts and partial deficiencies in other macrophage populations. These abnormalities could be corrected by the administration of exogenous M-CSF (Wiktor-Jedrzejczak et al. 1991; Umeda et al. 1996) and resolved spontaneously as the animals aged (Begg et al. 1993). Other deficient cell populations, which only responded minimally to exogenous M-CSF, included microglia in brain, synovial A cells and MOMA-1^+ or ER-TR9^+ macrophages. Their development seemed to be due either to M-CSF produced in situ or to expression of the M-CSF receptor (Wiktor-Jedrzejczak et al. 1991; Umeda et al. 1996). Although osteoclastic production was restored by M-CSF in op/op cell cultures, differentiation of these cells could not be induced (Hattersley et al. 1991). It seems that macrophages induced by M-CSF suppressed the differentiation to osteoclasts which is also induced by M-CSF (Hattersley et al. 1991).

Table 9. Modulation of M-CSF production

Cells	Primary stimulus	Augmenting factors	References	Inhibiting factors	References
monocytes/ macrophages	GM-CSF	IFN-γ M-CSF TNF-α	Gruber and Gerrard 1992		
	LPS	indomethacin	Lee et al. 1990	PGE, IL-4 dexamethasone	Lee et al. 1990 Hamilton 1994 Hamilton 1994
	TNF-α	IFN-γ	Oster et al. 1987	TNF-β	Oster et al. 1987
	anti-CD44; antiCD54; anti-LFA-3	LPS IL-1	Gruber et al. 1992 Gruber et al. 1992		
	general			HIV infection	Esser et al. 1996
smooth muscle cells	IL-1; TNF-α	cycloheximide	Filonzi et al. 1993		

Mice lacking both GM-CSF and M-CSF displayed the combined features corresponding to mice deficient in either factor alone: they had osteopetrosis, were toothless and had a characteristic alveolar-proteinosis which was more severe than that of only GM-CSF-deficient mice and was often fatal (Lieschke et al. 1994b). Some older mice with these mutations had polycythemia, their survival was reduced in comparison to mice deficient in either factor alone and they all had broncho- or lobar-pneumonia at death. These mice had circulating monocytes at levels comparable with those in M-CSF-deficient mice and their diseased lungs contained numerous phagocytically active macrophages, indicating that there are alternative mechanisms to ensure macrophage production and function in vivo.

M-CSF toxicity in humans is minimal and limited to reversible though sometimes dose-limiting thrombocytopenia and ophthalmological symptoms (Vial and Descotes 1995).

C
Receptors and Signal Transduction

M-CSF receptors are 165 kd proteins encoded by the c-fms proto-oncogene, which is also located on chromosome 5 (Cannistra and Griffin 1988). The number of receptors per cell is higher than for the other CSF at 3,000 to 16,000 per cell, though the degradation of the receptor complexes appears to be very rapid (Metcalf 1985). M-CSF molecules bound to cell-surface receptors are internalized and degraded in lysosomes (Horiguchi et al. 1986). The receptors have tyrosin kinase activity and their autophosphorylation has been noted after the binding of M-CSF (Metcalf 1985).

GM-CSF and IL-3 significantly decreased levels of M-CSF receptor and its mRNA in a murine myeloid precursor cell line, probably through reduction of the stability of the mRNA (Gliniak and Rohrschneider 1990; Rapoport 1992).

Introduction of the M-CSF receptor into hematopoietic cell lines of myeloid and T-lymphoid origin and addition of M-CSF to the medium led to the proliferation of only the myeloid cells, suggesting that cells of this origin, which do not normally express the receptor, still have an intact signal transduction pathway (von Ruden et al. 1991). But, although a murine hematopoietic cell line with this receptor proliferated clonally when stimulated by GM-CSF, IL-3 or M-CSF alone, the combination of M-CSF with either other factor strongly inhibited colony-formation, with loss of clonogenicity in affected cells accompanied by increased macrophage differentiation (Metcalf et al. 1992).

1.2 Role in Hematopoiesis

Of the four CSF discussed, M-CSF has the peculiarity of stimulating the proliferation of only a single type of leukocyte lineage, the monocyte/macrophage population. Furthermore, the effects on non-hematopoietic cells are limited to very few examples.

Human M-CSF selectively stimulated monocyte/macrophage colony-formation from normal human bone marrow in vitro, although the stimulus rhuM-CSF was more potent (Clark and Kamen 1987; Motoyoshi et al. 1989). Stimulation of bipotential cells by M-CSF throughout two to three cell divisions irreversibly commited the cells to form macrophage progeny, regardless of the CSF used subsequently to maintain proliferative stimulation (Metcalf 1985). Other less specific growth factors such as IL-3 and GM-CSF synergized with M-CSF and dramatically increased the number of monocytic colonies derived from human bone marrow (Caracciolo et al. 1987; Di Persio 1990; Herrmann et al. 1990). The combination of GM-CSF with M-CSF acting on murine bone marrow cells resulted in a significant decrease in the formation of certain macrophage colonies compared with that elicited by M-CSF alone (Metcalf and Nicola 1992). Pulmonary alveolar macrophages in vitro were induced to proliferate and develop into macrophage colonies by high doses of M-CSF and after a longer incubation time than with GM-CSF (Chen et al. 1988). However, this induction was greatly enhanced when a combination of M-CSF and a low dose of GM-CSF was employed (Chen et al. 1988). M-CSF, especially in combination with IL-3, stimulated a subpopulation of hematopoietic progenitors, likely osteoclast progenitors, that could give rise to TRAP$^+$ cells from monkey CD34 antigen positive bone marrow cells (Povolny and Lee 1993). However, another study found no TRAP$^+$ cells were formed in the absence of 1,25 dihydroxyvitamin D_3 or primary osteoblastic cells in a murine bone marrow culture system, but that the addition of M-CSF to these prerequisites enhanced the formation of TRAP$^+$ mononuclear cells better than did GM-CSF, G-CSF or IL-3 (Takahashi et al. 1991).

Both GM-CSF and IL-3 synergized with M-CSF in the induction of DNA synthesis in peripheral monocytes in vitro (Cheung and Hamilton 1992). The monocyte DNA synthesis due to M-CSF could be suppressed by IL-4, IFN-γ and TNF-α as well as cAMP-elevating agents BrcAMP, cholera toxin and PGE_2 (Cheung and Hamilton 1992).

M-CSF enhanced the proliferation of endothelial cells as well as macrophages from endotoxemic rats, which are more sensitive to CSF than cells from untreated rats (Feder and Laskin 1994). Only in the presence of IL-3 could M-CSF stimulate the proliferation of mast cells in vitro; alone it caused a decrease in their numbers (Rottem et al. 1994).

Injection of rmuM-CSF into mice, especially after pretreatment with human lactoferrin, increased the fraction of progenitors in active cell cycle in an in vivo study with normal mice (Broxmeyer et al. 1987). Intravenous administration of rmuM-CSF to mice pretreated with cyclophosphamide also increased the fraction of bone marrow progenitor cells in active cell cycle (Cannistra and Griffin 1988). In mice receiving a single s.c. dose of M-CSF, the number of F4/80^{+} alveolar macrophages, Kupffer cells and splenic macrophages increased with low doses of M-CSF, but did not alter in either of the groups when high doses of the factor were employed (Held et al. 1996). Single daily i.v. administration of rhuM-CSF to mice for four days led to a dose-dependent increase in circulating monocyte counts that was 10-fold at the highest dose tested as well as to an increase of the macrophage content of liver and peritoneal cavity (Herrmann et al. 1990). In contrast, multiple daily injections of the factor into mice suppressed progenitor cell cycling and failed to increase bone marrow, spleen, or blood cellularity (Chikkappa et al. 1989).

Recombinant human as well as purified urinary M-CSF administered to healthy volunteers caused an increase in the numbers of circulating monocytes without changing the bone marrow cellularity (Herrmann et al. 1990).

2 Effects on Granulocytes

No stimulating effects of M-CSF on either type of granulocytes have been documented so far: Neither rat nor human PMN responded to rhuM-CSF in respect to oxidative burst, chemotactic activity and adherence protein expression (Nathan 1989; Griffin et al. 1990; Wheeler et al. 1994). Also, M-CSF had no effect on the chemotactic activity (Yamaguchi et al. 1992a) or survival of basophils (Yamaguchi et al. 1992b).

3 Effects on Mononuclear Cells

3.1 Effects on the Functions of Monocytes/Macrophages

Of all the leukocyte lineages, M-CSF, in contrast to the other CSF discussed, apparently only affects the numerous functions of the monocyte/macrophage population.

A
Maturation of Monocytes to Macrophages

Blood monocytes differentiated in vitro to macrophages in the presence of M-CSF (Becker et al. 1987; Lopez et al. 1993).

B
Phagocytosis

The macrophage colonies resulting from incubation of pulmonary macrophages in vitro with M-CSF possessed pronounced FcR-mediated phagocytic activity (Chen et al. 1988). M-CSF enhanced the phagocytosis of opsonized heat-killed yeast (Bober et al. 1995b). Further, M-CSF was effective in preventing dexamethasone-induced suppression of monocyte anti-bacterial (*Staphylococcus aureus*) and anti-fungal (*Candida albicans*) phagocytic capacities (Bober et al. 1995b).

C
Oxidative Burst

M-CSF primed human monocytes and enhanced O_2^- release in response to the receptor-mediated agonists fMLP or ConA, but not to PMA which stimulates the cells independent of receptors. This effect was less pronounced than with GM-CSF and of similar extent as with IL-3 (Yuo et al. 1992). However, M-CSF had no effect on the ability of macrophages to exhibit a respiratory burst after *Listeria* infection in vitro (Denis and Gregg 1991).

D
Chemotaxis and Migration

Monocyte chemotaxis towards either the chemoattractants fMLP or LTB_4 was enhanced by M-CSF in culture (Bober et al. 1995b).

E
Tumor Cytotoxicity

M-CSF has been reported to stimulate the ADCC of normal human macrophages (Cannistra and Griffin 1988), especially in conjunction with a secondary signal such as IFN-γ (Baldwin et al. 1993). However, M-CSF failed to activate macrophages for effector activities toward fibrosarcoma, lymphoma or *Leishmania tropica* (Ralph et al. 1983), but pre-incubation of human peripheral blood monocytes with M-CSF markedly enhanced their ADCC toward a Burkitt lymphoma-derived cell line in a concentration dependent manner (Suzu et al. 1990). Pretreatment of peptone-elicited macrophages with M-CSF induced moderate killing and greatly stimulated lymphokine-induced killing of sarcoma cells (Ralph and Nakoinz 1987), but M-CSF did not stimulate freshly harvested exsudate macrophages to lyse these targets whether in the presence or absence of lymphokine activators. Therefore, M-CSF may be useful in combination with these activators in promoting resistance to cancer in mature mononuclear cells.

M-CSF further induced antibody-independent monocyte tumoricidal activity against the WEHI-164 murine fibrosarcoma cell line (Cannistra and Griffin 1988). Pre-incubation of human peripheral blood monocytes with M-CSF resulted in more effective killing activity towards K562, U937, Daudi, and HL60 cells than when the monocytes were pre-incubated with medium alone (Suzu et al. 1989). Anti-TNF antiserum partially blocked this tumoricidal activity augmented by M-CSF. The factor human peripheral blood monocytes produced when incubated with M-CSF and induced with LPS and PMA that was cytotoxic to L929 cells, was also identified as TNF-α (Warren and Ralph 1986). These results have been reproduced in vivo in studies in mice, where the sequential injection of rhuM-CSF and LPS showed maximal secretion of TNF-α (Sakurai et al. 1994). Neither M-CSF alone, nor LPS alone induced cytotoxic activity. Furthermore, pre-injection of M-CSF further enhanced the priming effect of IFN-γ for TNF production. As M-CSF treatment has been proven to prevent myelosuppression induced by murine IFN in vitro and in vivo, the combination of these two factors, which caused a synergistic priming effect on endogenous TNF production (Sakurai et al. 1994) may be effective in cancer therapy without severe myelosuppression.

A preliminary experiment where rhuM-CSF/ LPS treatment was applied in cancer therapy in mice with metastatic B16 melanoma, resulted in all mice having reduced or no visible tumor colonies in the liver, but lung metastatic tumor colonies were only weakly inhibited (Sakurai et al. 1994).

F
Synthesis of Mediators and Enzymes

M-CSF appears to induce little mediator release and only some mRNA synthesis by itself. M-CSF treatment of mouse bone marrow-derived macrophages led to a rapid and sustained increase in IL-1α and IL-1β mRNA as well as IL-1ra mRNA (Matsushime et al. 1991). Cycloheximide inhibited the M-CSF-induced IL-1α mRNA synthesis, but augmented IL-1β mRNA production and did not affect the induction of IL-1ra mRNA. Murine peritoneal macrophages treated with human M-CSF or L929-derived M-CSF did not reveal either PGE_2, IL-1 or IL-6 secretion (Strassmann et al. 1991). There was no increase in IL-1α mRNA or IL-6 mRNA levels in these cells. M-CSF induced murine bone marrow macrophages did not synthesize or release PGE_2, despite active PLA_2 (Shibata et al. 1994). In vitro, M-CSF alone, however, stimulated human mature monocytes prepared from the peripheral blood of healthy volunteers to produce GM-CSF and G-CSF protein, but neither IL-1 nor IFN-γ (Ishizaka et al. 1986; Motoyoshi et al. 1989) nor IL-1ra (Jenkins and Arend 1993). Murine peritoneal cells also produced G-CSF in response to M-CSF (Metcalf and Nicola 1985).

M-CSF enhanced the mRNA and protein levels PAI 1 and 2 (Hamilton et al. 1993). PAI may modulate the effects of CSF on monocyte urokinase-type PA activity at sites of inflammation and tissue remodeling (Hamilton et al. 1993). However, there was also evidence that M-CSF directly stimulated the production of a u-PA activity in both murine bone marrow-derived and peritoneal macrophages (Hamilton et al. 1991). This increase in u-PA activity could be abrogated by dexamethasone, PGE_2 and cholera toxin (Hamilton et al. 1991). M-CSF also stimulated the basal production of complement factor C3 by monocytes in vitro, though neither basal factor B nor LPS-stimulated production of either of the factors was affected by M-CSF (Høgåsen et al. 1993).

M-CSF and LPS synergized in vitro and induced a murine thymocyte cell line to express higher levels of mRNA for IL-1α, IL-1β, TNF-α and IL-6 and to cause the release of more bioactivity than macrophages treated with LPS alone (Evans et al. 1992). Human peripheral blood monocytes incubated with M-CSF and induced with LPS and PMA produced significantly elevated levels of TNF-α and colony-stimulating activity (Warren and Ralph 1986). M-CSF-treated monocytes induced with poly-I C secreted increased levels of IFN-γ (Warren and Ralph 1986). Murine peritoneal macrophages, however, incubated with both LPS and M-CSF produced less IL-1 bioactivity and IL-1ra than cells incubated with only LPS (Strassmann et al. 1991). M-CSF also had no effect on the production of IL-1ra protein in human monocytes stimulated with LPS or cultured on adherent IgG (Jenkins and Arend 1993). In murine macrophages, M-CSF synergized with LPS and IFN-γ to induce nitric oxide production (Feder and Laskin 1994).

Injection of mice with M-CSF resulted in dose-dependent elevated TNF-α and IL-6 levels in the bronchoalveolar lavage fluid (Held et al. 1996). Also, the sequential injection of rhuM-CSF and LPS into mice caused secretion of TNF-α into the serum (Sakurai et al. 1994). Furthermore, pre-injection of M-

CSF also enhanced the priming effect of IFN-γ for TNF production (Sakurai et al. 1994).

Taken together, M-CSF appears to exert effects similar to GM-CSF, i.e. it induces little mediator release on its own, but primes leukocytes for increased release of pro-inflammatory factors.

G Expression of Surface Molecules

In contrast to GM-CSF and IL-3, M-CSF partially inhibited the expression of GM-CSF receptors on peritoneal exsudate macrophages (Fan et al. 1992). High concentrations of M-CSF downregulated M-CSF receptor mRNA expression in immature progenitors and monocytes derived from bone marrow $CD34^+$ cells in culture (Panterne et al. 1996). The expression of M-CSF-induced CD23 was reduced by TNF-α by suppression of the mRNA and by enhancement of the release of the soluble receptor (Hashimoto et al. 1997). M-CSF failed to augment the expression of CD54 or HLA class I or II in human monocytes (Sadeghi et al. 1992). Instead, M-CSF suppressed the basal levels of Ia gene expression in bone marrow-derived macrophages, and also inhibited its induction by IFN-γ or GM-CSF (Fischer et al. 1988; Willman et al. 1989).

The expression of Ia antigen and CD11b/CD18 complex increased on the splenic macrophages of mice treated in vivo with M-CSF, but was not affected in their alveolar macrophages and Kupffer cells (Held et al. 1996).

H Microbial Killing

M-CSF made human monocytes more vulnerable towards two virulent strains of *Mycobacterium avium* and stimulated the extracellular growth of this parasite in vitro in tissue culture medium (Denis 1991b). Macrophages pre-treated with M-CSF were also more permissive for the growth of *Listeria monocytogenes* (Denis and Gregg 1991). Treatment of macrophages previously infected with *Leishmania mexicana amazonensis* with M-CSF caused a significant dose-dependent reduction in intracellular parasites in vitro. This effect was augmented by the addition of IFN-γ (Ho et al. 1990). M-CSF-conditioned medium failed to activate macrophages for effector activities toward *Leishmania tropica* (Ralph et al. 1983). At relatively high concentrations, M-CSF enhanced monocyte toxicity toward *Candida albicans*, but to a smaller degree than IL-3 or GM-CSF (Wang et al. 1989).

I
Other

Human monocyte-derived macrophages cultured in medium containing M-CSF were more susceptible to infection with HIV-1: The frequency with which the cells became infected, the level of HIV mRNA expressed per infected cell and the level of proviral DNA expressed per infected culture were all increased by M-CSF (Gruber et al. 1995). Infected cells maintained in the absence of exogenous M-CSF produced this cytokine at high levels (Gruber et al. 1995), but the pro-inflammatory cytokines were not produced. The endogenous M-CSF production may contribute to the survival of HIV-infected macrophages and enable them to function as a reservoir for HIV and facilitate the spread of the virus in vivo.

3.2
Effects on the Functions of Lymphocytes

To the author's knowledge, no effects of M-CSF on the functions of lymphocytes, which express no receptors for M-CSF, have been documented so far.

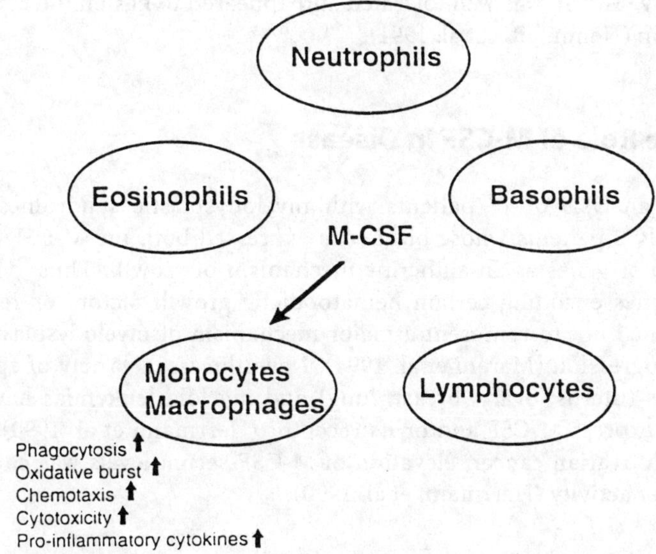

Fig. 4. Pertinent immunomodulatory effects of M-CSF on different leukocyte populations. The diagramm summarizes the predominant effects of M-CSF on immune functions of different types of leukocytes

4 Effects on the Functions of Other Cells

The effects of M-CSF discussed above are displayed in Fig. 4. As with the other CSF, some effects of M-CSF in other cells have been recorded. In endothelial cells, M-CSF synergized with LPS and IFN-γ to induce nitric oxide production (Feder and Laskin 1994). M-CSF also induced the expression and synthesis of the insulin-like growth factor-1 (IGF-1) in murine bone marrow cells (Kelley et al. 1996).

5 Infection

M-CSF administration to mice protected them from a subsequent lethal challenge with *Candida albicans*, demonstrated by increased survival times and a reduction in recovery of viable *C. albicans* from various organs (Cenci et al. 1991).

In a phase 1 uncontrolled trial, bone marrow transplantation patients with severe fungal infections were given M-CSF in conjunction with antifungal therapy. M-CSF was well tolerated and appeared to hasten the resolution of infection (Nemunaitis et al. 1991).

6 Possible Role of M-CSF in Disease

Genomic analysis of 17 patients with myelodysplastic syndromes determined only 2 patients whose blast cells expressed both the M-CSF and M-CSF receptor genes as an autocrine mechanism of growth. Thus, deregulation of genes encoding certain hematopoietic growth factors or receptors were deemed not to represent a major mechanism of myelodysplastic syndrome progression (Mareni et al. 1994). Nevertheless, a variety of epithelial neoplasms (uterus, ovary, breast, lung) and myeloid leukemias have been shown to express M-CSF and/or its receptor (Herrmann et al. 1990). In patients with ovarian cancer, elevation of M-CSF serum levels was correlated with cancer activity (Herrmann et al. 1990).

7 Potential Clinical Applications

When 46 patients with gynecologic malignancies in a randomized controlled study received urinary M-CSF in chemotherapy cycles, the average period of neutropenia was significantly shortened and the average nadir of platelets increased significantly in comparison to identical cycles without exogenous M-CSF in the same patients (Herrmann et al. 1990).

Clinical trials in Japan have demonstrated that M-CSF can relieve the myelosuppression associated with a number of diseases including chronic childhood neutropenia and myelodysplastic syndromes (Garnick and Reilly 1989). Other prospective clinical roles that have been proposed for M-CSF include acting as a promoter of monocyte-mediated tumor cell cytotoxicity and as a terminal differentiation inducer for AML cells (Whetton 1990).

IV IL-3

1 General Information

1.1 Molecular Biology and Endogenous Production

A The Molecule Interleukin-3

This protein was originally described in 1963 as a mast cell growth factor in thymic cell cultures. Later, IL-3 was found to be constitutively produced by myelomonocytic WEHI 3 cells and was biochemically characterized and purified from the culture supernatant of this cell line in 1982 (Frendl 1992). The cDNA clones of murine IL-3 were decoded in 1984, whilst the corresponding human sequence remained unknown until 1986 (Hamblin 1988).

The IL-3 gene, made up of 5 exons separated by one large and three small introns (Cannistra and Griffin 1988), is located together with GM-CSF, M-CSF, M-CSF-R, IL-4 and IL-5 on chromosome 5 in humans. IL-3 and GM-CSF are evolutionarily related, have some similarity in structure (Yang et al. 1988) and their expression is coordinately regulated to a certain extent (Frendl 1992). Human IL-3 consists of 152 amino acids with a signal sequence of 19 amino acids (Hamblin 1988) and exhibits one disulphide bond. The apparent molecular weight of IL-3 ranges from 14 to 30 kd, although the expected size of the mature polypeptide is only 14 to 15 kd. This heterogeneity in size results from variable degrees of glycosylation at two asparagine residues (Cannistra and Griffin 1988).

The homology between human and murine IL-3 is 29% at amino acid level, which explains why these two proteins are not cross-reactive (Cannistra and Griffin 1988). There is a 93% homology between human and gibbon IL-3 though, so these two factors should have identical biological activities (Clark and Kamen 1987).

B
Endogenous Production of IL-3

IL-3 is produced by activated T-cells (Yang et al. 1986; Otsuka et al. 1988), mast cells (Wodnar-Filopowicz et al. 1989; Plaut et al. 1989), neutrophils (Cassatella 1995), eosinophils (Kita et al. 1991), epidermal keratinocytes and thymic epithelial cells in vitro (Frendl 1992). Furthermore, IL-3 was found to be expressed in the developing fetal thymus, in tissues that had undergone syngeneic or allogeneic transplantation (Frendl 1992) and during allergen-induced late-phase cutaneous reactions in atopic patients (Kay et al. 1991). IL-3 is the only cytokine produced during syngeneic mixed lymphocyte reactions (Suzuki et al. 1986), indicating a role in the interaction of syngeneic antigen-presenting cells and T-cells.

Purified T-lymphocytes produced IL-3 in response to PHA plus PMA and IL-2 (Oster et al. 1989c). The inductive effects of ConA or ConA plus PMA on IL-3 production by T-cells were abrogated by factors which activate the cAMP signaling pathway, such as dibutyryl-cAMP, PGE_2 or isoproterenol (Borger et al. 1996). PBMC could be stimulated by the T-cell mitogen PHA (McHugh et al. 1996) or PMA (Oster et al. 1989c; Cassatella 1995) to produce IL-3. IL-3 secretion by T-cells is probably not involved in the maintenance of baseline hematopoiesis, as no stable mRNA was produced without a secondary signal in vitro (Oster et al. 1989c). It is more likely that T-lymphocytes are activated by antigen to produce IL-3 at peripheral sites of infection so as to amplify the effector function of local granulocytes and monocytes (Cannistra and Griffin 1988). Both Th_1 and Th_2 type $CD4^+$ clones produced substantial amounts of IL-3 upon activation in vitro (Frendl et al. 1990) and it has been determined that the expression of IL-3 and GM-CSF are differentially regulated in activated $CD8^+$ and $CD4^+$ T-cell clones (Fitzpatrick and Kelso 1995).

Table 10. Factors affecting the production of the IL-3 protein

Stimulus type	Stimulus	Cell type	Results	References
Constitutive	none	T cells	–	Oster et al. 1989c
Other	PHA PMA PMA + PHA	PBMC PBMC T cells	+ + +	Oster et al. 1989c Oster et al. 1989c Oster et al. 1989c
	Con A Con A + PMA	T cells T cells	+ +	Borger et al. 1996 Borger et al. 1996
	IL-2	T cells	+	Oster et al. 1989c
	ionomycin	PMN eosinophils	+ +	Casatella 1995 Kita et al. 1991
	IgE	mast cells	+	Wodnar Filipowicz et al. 1989

Key: + : induction of IL-3 production and secretion
\- : no effect on the production of IL-3

Although PMA-stimulated mononuclear cells made greater quantities of IL-3 than PMN stimulated with ionomycin under similar conditions on a single-cell basis, one must consider that granulocytes constitute the majority of infiltrating cells in inflamed tissues (Cassatella 1995). Human peripheral blood eosinophils were stimulated to release IL-3 in vitro by ionomycin; an effect which was inhibited by cyclosporin A (Kita et al. 1991). IgE-receptor-mediated activation of murine mast cells in vitro, as happens in allergic diseases in vivo, resulted in the production of IL-3 (Wodnar-Filopowicz et al. 1989).

Table 10 summarizes the activity of various factors in promoting the production of IL-3.

Serum IL-3

There have been virtually no situations in mice that have resulted in detectable circulating levels of IL-3, even under the conditions that result in IL-3 production in vitro (Nicola 1989). In mice infected with *Listeria monocytogenes*, no IL-3 was detectable in the serum (Cheers et al. 1988).

Toxicity

No major side effects of IL-3 administration have been observed in non-human primates, other than some skin lesions characterized as urticaria, which may be due to basophil activation (Whetton 1990). Toxicities reported with IL-3 in humans include flu-like and constitutional symptoms, severe headache or skin rash (Vial and Descotes 1995).

C
Receptors and Signal Transduction

The high-affinity IL-3 receptor is expressed on eosinophils, basophils and monocytes (Valent et al. 1989a; Lopez et al. 1989; Elliot et al. 1989), but not on mature neutrophils (Weisbart et al. 1989). The subclass II GM-CSF receptor, which is present on AML cells and normal monocytes (Oster et al. 1991), can bind either GM-CSF or IL-3 (Frendl 1992) which compete with an equally high affinity. Cross-competition between GM-CSF and IL-3 for binding has also been confirmed in eosinophils (Lopez et al. 1989) and basophils (Lopez et al. 1990). Cloning of the IL-3 receptor yielded a 140 kD protein that can bind IL-3 with low affinity upon transfection into fibroblasts (Itoh et al. 1990). A second membrane protein has been found that does not bind IL-3 itself, but its expression is co-regulated with the described 140 kD protein (Frendl 1992). However, neither of these components possess tyrosine kinase activity or high-affinity binding characteristics, indicating the involvement of additional receptor components (Frendl 1992). It therefore seems that the IL-3 receptor, similar to the GM-CSF receptor, is composed of two chains, a ligand-specific α-chain, which binds the cognate ligand with low affinity and a common β-chain which, although it cannot bind the ligand itself, confers high affinity binding when co-transfected with the α-chains (Lopez et al. 1992). The sharing of the β-chain may explain the cross-competition between the two factors.

The primary event in IL-3-induced cellular activation seems to be protein tyrosine phosphorylation, especially of the MAP kinases p42 and p44 (Okuda et al. 1992), independent of protein kinase C (PKC) (Frendl 1992). The role of pertussis toxin-sensitive G-proteins in the IL-3-mediated cellular activation process has been specifically demonstrated and was found to affect primarily the cells of the macrophage lineage (Frendl 1992).

1.2 Role in Hematopoiesis

In vitro and in vivo, IL-3 primarily stimulates the proliferation of early progenitors and supports the specific actions of other endogenous or exogenous hematopoietic growth factors, especially M-CSF, G-CSF and stem cell factor.

The term "multi-CSF" was coined because the growth of various multipotent progenitors (Spivak et al. 1985; Cannistra and Griffin 1988; Sieff et al. 1987b) is supported by IL-3, as is that of progenitors with a more restricted commitment status such as granulocyte/monocyte, eosinophil and megakaryocyte colony-forming units (Sieff et al. 1987b). IL-3 did not trigger cell cycling of dormant stem cells, but instead supported the proliferation of multipotential progenitors after they exited the G_0-phase (Ogawa 1993). IL-3 was able to induce growth and differentiation of human basophils in bone marrow suspension cultures (Valent et al. 1989b) and also induced the proliferation of persisting cells in a dose-dependent manner (Ihle et al. 1983).

IL-3 further stimulated the growth of mast cells from bone marrow cultures alone (Ihle et al. 1983; Rottem et al. 1994) or in combination with stem cell factor and, to a lesser extent, with IL-4 and IL-9 (Ihle et al. 1983; Rottem et al. 1994). The addition of M-CSF, GM-CSF or IFN-γ, on the other hand, decreased the number of mast cells in these experiments (Rottem et al. 1994). IL-3, in synergy with M-CSF, stimulated a subpopulation of hematopoietic progenitors, probably osteoclasts, that could give rise to TRAP$^+$ cells (Povolny and Lee 1993). The addition of GM-CSF to cultures grown in the presence of IL-3 did not result in enhanced colony-formation, suggesting that most if not all GM-CSF progenitors also respond to IL-3 (Clark and Kamen 1987). In contrast, addition of G-CSF to IL-3-containing cultures resulted in substantially greater numbers of neutrophil colonies than with IL-3 alone, without depressing colony-formation of other cell types (Clark and Kamen 1987; Weisbart et al. 1989).

Infusion of IL-3 into mice led to splenomegaly with a concomitant increase in the numbers of splenic myeloid progenitor cells present (Whetton 1990). Mice pretreated with sublethal irradiation showed a tenfold increase of bone marrow progenitor cells to near normal levels during a 7 day treatment with IL-3 (Kindler et al. 1986). Single injections of IL-3 increased the fraction of hematopoietic progenitors in active cell cycle in an in vivo study with normal mice (Broxmeyer et al. 1987). Here, pretreatment with human lactoferrin made the cells more sensitive toward IL-3. In another study, mice receiving IL-3 i.p. over six days showed a tenfold elevation of blood eosinophil levels and a threefold elevation of peripheral neutrophil and monocyte counts (Metcalf et al. 1986). Their spleens were enlarged by an accumulation of mast cells, maturing granulocytes, eosinophils, nucleated erythroid cells and megakaryocytes. A significant rise in intraperitoneal macrophage levels was also noted (Metcalf et al. 1986). IL-3 and G-CSF administered concurrently as an i.v. injection in rats induced a peripheral neutrophilia that was approximately additive in comparison to the neutrophilia induced by the factors individually (Ulich et al. 1990b). Daily injection of IL-3 plus G-CSF also caused a significant decrease in erythroid, lymphoid, and eosinophilic marrow precursors in rats, despite the fact that IL-3 alone induced a significant erythroid hyperplasia (Ulich et al. 1990b).

During the second week of s.c. IL-3 administration rhesus monkeys responded with a two- to threefold increase of WBCs caused by a dose-dependent elevation of basophils (up to 40% of WBCs) and eosinophils (Mayer et al. 1989).

No information on IL-3 deficiency, excessive IL-3 production or toxicities related to the use of IL-3 in humans is available at present.

2 Effects on Granulocytes

2.1 Effects on the Functions of Neutrophilic Granulocytes

Freshly isolated mature neutrophils have no IL-3 receptors and are therefore not responsive to IL-3 (Smith et al. 1995). However, incubation with GM-CSF induces the expression of the α-subunit of the IL-3-R. Subsequent addition of IL-3 results in the expression of HLA-DR which is greater than with GM-CSF alone (Smith et al. 1995).

Stimulation of PMN with high concentrations (1000 U/ml) of IL-3 resulted in adhesion of the cells to a plastic layer and development of long protrusions (Zeck-Kapp et al. 1989). In contrast to GM-CSF, a lower number of intracytoplasmic vesicles was detected (Zeck-Kapp et al. 1989). Production of H_2O_2 was observed at the granulocytes' outer surface and in the luminal part of the vesicles (Zeck-Kapp et al. 1989).

2.2
Effects on the Functions of Eosinophilic Granulocytes

A
Phagocytosis

IL-3 increases the phagocytotic activity of human eosinophils toward *Candida albicans* (Fabian et al. 1992), but does not enhance the killing of this fungus.

B
Chemotaxis and Migration

Eosinophils purified from patients with hypereosinophilic syndrome became adherent in the presence of IL-3 (Tai et al. 1990). Pre-incubation of eosinophils with picomolar concentration of IL-3 induced a chemotactic response toward IL-8 and fMLP and enhanced PAF-induced chemotaxis (Warringa et al. 1991). Nanomolar concentrations of IL-3 inhibited the C5a-induced chemotaxis, unless the cells were washed after pre-incubation (Warringa et al. 1991).

C
Synthesis of Mediators

IL-3 enhanced the generation of LTC_4 induced by calcium ionophore A23187 in human eosinophils (Fabian et al. 1992).

D
Degranulation

IL-3-activated eosinophils transformed the storage form of eosinophil cationic protein (ECP) into the secreted form in vitro, but this was only released when the cells were exposed to secretory stimuli, such as sepharose coated with C3b or sepharose-activated whole autologous serum in greater amounts than with the stimuli alone (Tai et al. 1990). The release of arylsulphatase and ß-glucuronidase from specific and small granules of human eosinophils was stimulated by IL-3 in vitro (Fabian et al. 1992).

E
Expression of Surface Molecules

IL-3 increased PAF receptor levels on human eosinophils in vitro (Kishimoto et al. 1996).

F
Microbial Killing

The killing of *Staphylococcus*, but not of *Candida* by human eosinophils was increased by IL-3 (Fabian et al. 1992).

G
Other

IL-3 prolonged the lifespan of blood eosinophils in a dose-dependent manner by preventing apoptosis and inducing the expression of the activation forms of eosinophil ribonucleases (Tai et al. 1990 and 1991).

2.3
Effects on the Functions of Basophilic Granulocytes

A
Histamine Release

IL-3 has been described as an effective, direct, time and temperature dependent histamine releasing factor acting on basophils from allergic and, less frequently, from normal subjects (Haak-Frendscho et al. 1988; Miadonna et al. 1993). However, in another study, only relatively high doses of IL-3 directly caused a release of small amounts of histamine is some allergic donors' cells (Alam et al. 1989).

IL-3 enhanced the release of histamine on stimulation via the IgE receptor or with fMLP or ionophore A23187 by basophils in vitro in a rapid, temperature-dependent manner, presumably along the same pathway as GM-CSF, as they both reached a similar plateau value and showed no additive effects when combined (Hirai et al. 1988; Miadonna et al. 1993). However, a different study determined that IL-3 had a greater potency in stimulating IgE-mediated histamine release than GM-CSF and that it effected the release of greater amounts of histamine than GM-CSF (Lopez et al. 1990). IL-3 could also mediate the release of histamine when present at low concentrations when incubated with basophils in the presence of D_2O in some allergic donors (Alam et al. 1989). Pretreatment of basophils with IL-3 makes them responsive to very low concentrations of C3a, resulting in a rapid release of large amounts of histamine and also in the generation of leukotrienes (Bischoff et al. 1990). This phenomenon could be of relevance in various inflammatory processes, especially hypersensitivity reactions (Bischoff et al. 1990).

Consistent with these in vitro data, IL-3 induced a dose-dependent increase of histamine (up to 700-fold above normal values) in the blood of monkeys treated with IL-3 (Mayer et al. 1989).

B
Expression of Surface Molecules

CD 11b/CD18 complex expression on peripheral blood basophils was induced by IL-3 in vitro (Bochner et al. 1990).

C
Adhesion and Chemotaxis

IL-3 promoted basophil adherence to vascular endothelium (Bochner et al. 1990). IL-3 induced chemotacic activity in human basophils, but whether the induced migration form is chemotaxis or chemokinesis is still controversial (Tanimoto et al. 1992; Yamaguchi et al. 1992a).

D
Other

IL-3 maintained numbers of viable human basophils in culture (67% at day 7; 11% in controls) (Yamaguchi et al. 1992b).

3
Effects on Mononuclear Cells

3.1
Effects on the functions of monocytes/macrophages

A
Maturation of Monocytes to Macrophages

Monocytes differentiated to macrophages in the presence of IL-3 in vitro (Lopez et al. 1993). IL-3 induced morphologic changes in macrophages, including increased spreading, vacuolation and number of cytoplasmic processes (Gillan et al. 1993).

B
Phagocytosis

Increased phagocytosis of opsonized yeast in IL-3-activated, bone marrow-derived and peritoneal macrophages in vitro were reported (Crapper et al. 1985). When mice received IL-3 i.p. for a period of 6 days, an enhancement of macrophage phagocytic activity was noted (Metcalf et al. 1986).

C
Oxidative Burst

IL-3 primed monocytes to increase the production of ROS in response to the secondary stimuli fMLP and ConA, but not in response to PMA which binds no receptor. The effect was smaller than with GM-CSF, but similar to that achieved with M-CSF (Yuo et al. 1992).

D
Tumor Cytotoxicity

IL-3 enhanced monocyte killing of WEHI 164 fibrosarcoma cells by a TNF-dependent mechanism in response to a second stimulatory event, e.g. endotoxin in vitro (Cannistra et al. 1988). However, IL-3 was unable to induce tumoricidal activity of macrophages toward P815 target cells and could not modulate the inductive effect of IFN-γ in this regard (Frendl 1992).

E
Synthesis of Mediators and Enzymes

Monocytes responded to IL-3 with the expression of the pro-inflammatory cytokines IL-1β, IL-6, IL-8 and TNFα, the autoregulatory IL-1ra and the M-CSF (Cluitmans et al. 1993) as well as G-CSF (Oster et al. 1989b). Of these only IL-8 (Cannistra et al. 1988; Takahashi et al. 1993), IL-1ra (Jenkins and Arend 1993; Jenkins and Arend 1993) and G-CSF (Metcalf and Nicola 1985; Oster et al. 1989b) were translated into protein and secreted directly. In contrast, another report stated that IL-3 decreased the transcription rate of the IL-1β gene (Oster et al. 1992).

M-CSF-induced accumulation of IL-1β mRNA was enhanced by IL-3 via unknown posttranscriptional means that may relate to an increased expression of the M-CSF receptor mRNA (Oster et al. 1992). IL-3 seemed to synergize with GM-CSF in the induction of G-CSF in peripheral blood monocytes (Oster et al. 1989b). IL-3, in conjunction with IFN-γ, stimulated the production of TNF by monocytes, though, whether the production of IL-1 is also stimulated is controversial (Cannistra et al. 1988; Hart et al. 1990; Frendl et al. 1990). IL-3 was also found to contribute synergistically to LPS-induction of increased TNF-α (Cannistra et al. 1988; Hart et al. 1990; Frendl et al. 1990), IL-1 (Hart et al. 1990; Frendl et al. 1990) and IL-6 activities (Frendl 1992) by murine and human macrophage cell lines.

IL-3 directly stimulated the production of a u-PA in both murine bone marrow-derived and peritoneal macrophages (Hamilton et al. 1991). The increase in u-PA activity was abrogated by dexamethasone, PGE_2 and cholera toxin (Hamilton et al. 1991).

Taken together, IL-3 appears to favor and support pro-inflammatory mediator release.

F
Expression of Surface Molecules

High, but not low, doses of IL-3 resulted in the downmodulation of the M-CSF receptor in immature progenitors and monocytes in culture (Gliniak and Rohrschneider 1990; Panterne et al. 1996). IL-3 induced the upregulation of GM-CSF receptors on peritoneal exsudate macrophages, an effect which was abrogated by cycloheximide, a protein-synthesis inhibitor (Fan et al. 1992), and also induced the expression of the IL-1 receptor on human and murine bone marrow cells as well as lymphoid and myeloid progenitor cell lines (Frendl 1992).

Although one report described IL-3 as having no effect on the gene expression of the murine MHC component Ia in macrophages from bone marrow cultures (Willman et al. 1989), others have determined that IL-3 is able to induce the expression of class II MHC antigens in murine peritoneal exsudate cells (Frendl and Beller 1990). IL-3 and IFN-γ probably work via different mechanisms in this respect, as induction by IL-3 was delayed in comparison and the combination of IL-3 with IFN-γ or low doses of LPS produced a synergistic effect, but high doses of LPS inhibited the induction of Ia expression by IL-3 (Frendl and Beller 1990; Frendl 1992). IL-3 plus IL-4 induced monocyte expression of CD1, a molecule which presents bacterial antigens to certain T-lymphocytes (Thomssen et al. 1996). The stimulatory effect of IL-3 could be countered by IL-10.

IL-3 also had the potential to induce CD11a/CD18 expression with kinetics similar to IFN-γ or LPS in murine peritoneal exsudate cells (Frendl and

Beller 1990). Combination of IL-3 with IFN-γ had an additive effect in this respect, but neither a synergistic nor an additive effect could be established with the combination of IL-3 with LPS, already a potent inducer of CD11a/CD18 (Frendl 1992). IL-3 induced the expression of CD23 in monocyte cultures, an effect which was abrogated by the addition of TNF-α (Hashimoto et al. 1997).

G
Microbial Killing

The treatment of human monocytes with IL-3 led to increased permissiveness of these cells for two virulent strains of *Mycobacterium avium* and also dramatically increased extracellular *M. avium* growth in vitro in tissue-culture medium (Denis 1991b). IL-3 made macrophages more vulnerable to *Listeria monocytogenes* and had no effect on the oxidative burst capacity of macrophages after *Listeria* infection (Denis and Gregg 1991).

In contrast, IL-3 effectively enhanced human monocyte-mediated anticandidal activity in vitro by fresh as well as aged cells, even at very low concentrations (Wang et al. 1989).

H
Other

IL-3 was found to be able to sensitize human peripheral blood monocytes to lysis by autologous LAK cells (Djeu et al. 1989).

3.2
Effects on the Functions of Lymphocytes

A
Migration

Potent, temperature-dependent stimulation of human lymphocyte migration was observed in response to IL-3 in vitro; this effect was abolished in the presence of cytochalasin B (Bacon et al. 1990). IL-3-desensitized cells could no longer be stimulated to migrate by either IL-3 or IL-4, suggesting that the post-receptor signal transduction mechanism is the same for these two factors (Bacon et al. 1990).

B
Mediator Synthesis

PMA-differentiated AML-193 leukemic cells were stimulated to produce IL-1ra by IL-3 (Kindler et al. 1990).

4
Effects on the Functions of Other Cells

Although IL-3 acts mainly on leukocytes as summarized in Fig. 5, some effects on other cells have also been characterized. Mouse bone marrow cells cultured in the presence of IL-3 released PGE_2 and LTB_4 when stimulated with calcium ionophore A23187, but not when PMA was used (Shibata et al. 1990). Neither these cells nor IL-3-dependent cell lines released significant amounts of PGE_2 when stimulated with IL-3 alone, although translocation of protein kinase C to the membrane fraction was induced (Shibata et al. 1990), indicating that elevated cellular Ca^{2+} is required and PKC activation alone is an insufficient stimulus. IL-3 induced a marked increase in the histamine synthesis of normal C57BL/6 bone marrow cells (Ihle et al. 1983).

IL-3 induced dose-dependent IL-2 receptor (CD25) expression on early myeloid cells in normal human bone marrow, an effect which requires pro-

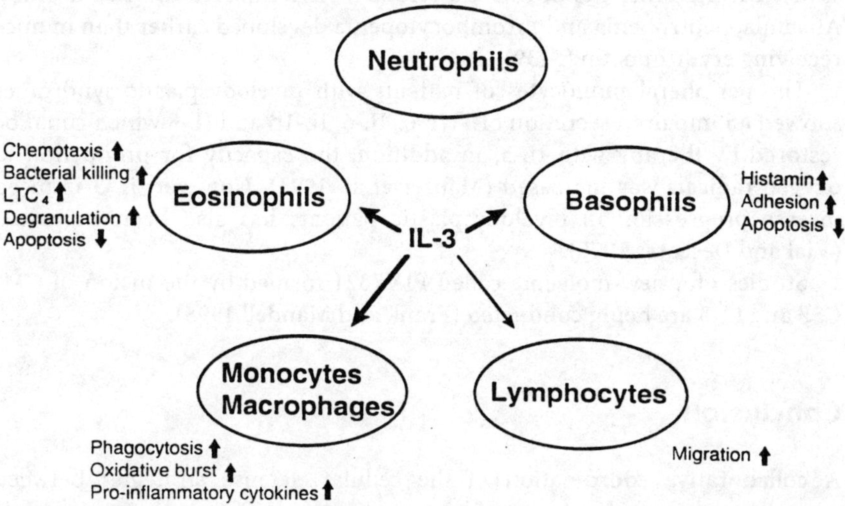

Fig. 5. Pertinent immunomodulatory effects of IL-3 on different leukocyte populations. The diagramm summarizes the predominant effects of IL-3 on immune functions of different types of leukocytes

tein synthesis as it was inhibited by cycloheximide (Gazzola et al. 1992). Similar to M-CSF, IL-3 induced the expression and synthesis of the IGF-1 in murine bone marrow cells (Kelley et al. 1996).

5
Role of IL-3 in Disease

IL-3, as well as GM-CSF, played a part in the macrophage infiltration of the white pulp of the spleen and brain in an experimental malaria model (Grau et al. 1988). Anti-IL-3 and anti-GM-CSF antibodies were beneficial in the prevention of the neurological symptoms of malaria infection in this experiment.

An IgG auto-antibody mimicking the effect of IL-3 has been found in the sera of a stem of autoimmune-prone mice (Ohta et al. 1988).

6
Potential Clinical Applications

Although IL-3, like GM-CSF, augmented HIV expression in monocytes, it did not interfere with the anti-retroviral activity of zidovudine and counteracted the myelosuppressive side effects of this drug (Schuitemaker et al. 1990). On the other hand, IL-3 exacerbated AZT-induced toxicity in mice: Anaemia, neutropenia and thrombocytopenia developed earlier than in mice receiving erythropoetin (S139).

The peripheral monocytes of patients with myelodysplastic syndromes showed an impaired secretion of TNF-α, IL-6, IL-1β and IL-8 which could be restored by therapy with IL-3; in addition, the capacity for production of oxygen radicals was increased (Maurer et al. 1993). However, IL-3-induced disease progression in myelodysplastic patients has also been suggested (Vial and Descotes 1995).

Studies of a new molecule called PIXY321 formed by the fusion of GM-CSF and IL-3 are being conducted (Frank and Mandell 1995).

Conclusion

A collaborative coordination of the cellular defense strategies between macrophages, granulocytes and the various subsets of lymphocytes is mandatory for the host to limit tissue damage through uncontrolled immune responses. The non-specific immune response mobilizes macrophages, neutrophilic granulocytes (PMN), natural killer cells and cytotoxic lymphocytes. The different CSF strongly contribute not only to the emergency formation of these cells, but also mature peripheral cells and modulate their functions.

Increased hematopoiesis enables continuous recruitment of leukocytes over the entire time span of an infectious disorder. An increased consumption of leukocytes to eradicate infectious material therefore requires the peripheral production of hematopoietic growth factors controlled from the site of infection as an efficient mechanism to guarantee the supply of leukocytes. Notably, the four CSF covered in this review are formed at the focus of infection and inflammation. In addition to changing the average maturity of peripheral cells by preferential release of premature forms of leukocytes from the bone marrow, hematopoietic growth factors can also promote the differentiation process of peripheral white cells.

During permanent overactivation of the immune system, the dilemma arises that any signal recruiting more leukocytes and activating them represents a source of potential danger. It therefore makes teleological sense to focus the activity of augmentation and activation of a given growth factor on a single leukocyte population with constraints of activity directed towards other leukocytes. This type of hierarchic regulation provides coordinated termination of activities and thereby avoids an overshoot of the inflammatory response. This might explain why CSF often exert opposite effects on different leukocyte subpopulations.

G-CSF differs from other CSF (perhaps with exception of M-CSF) with regard to the amount of cytokine formed: Serum levels can reach several hundred ng/ml and remain elevated over a long time span. This might explain how G-CSF, in contrast to the other CSF, transmits very pronounced anti-inflammatory activities, restricting the overall defense reaction.

In concordance with their assigned endogenous functions, CSF, when given to laboratory animals, volunteers or patients in the form of exogenous pure compounds, act as biological response modifiers by interfering with the body's humoral signaling system. Therefore, the therapeutic use of hematopoietic growth factors has potentials beyond alterating absolute as well as relative numbers of leukocytes. From the lessons learnt in basic research which assign each CSF with a specific role in the orchestration of the immune system, promising indications are expected to be identified. In addition, the identification of patient populations with an absolute or relative lack of production of the CSF will give a rationale for their therapeutic substitution.

G-CSF and GM-CSF have been used in close to 2 million patients since their introduction into the clinics in the late eighties. Over all, they have proven to be safe drugs and even broader use is restricted primarily by cost/efficacy considerations. In light of the emerging new findings of promising new indications, an expansion in the clinical use of these factors can be predicted easily.

In this sense, it is impressing to recall the prophecy of George Bernhard Shaw, who wrote in 'The Doctor's Dilemma' in 1904:

'There is at the bottom only one genuine scientifc treatment of all diseases: Stimulate the phagocyte – drugs are delusion!'

Acknowledgment. This paper is part of a habilitation thesis of the University of Konstanz conducted in the Department of Biochemical Pharmacology under the guidance of Prof. Albrecht Wendel. The invaluable help of Sonja von Aulock in preparing the manuscript is gratefully appreciated. The author is indebted to Dr. MaryAnn Foote and Dr. Albrecht Wendel for carefully reading the text.

References

Abraham E, Stevens P (1992) Effects of granulocyte colony-stimulating factor in modifying mortality from Pseudomonas aeruginosa pneumonia after hemorrhage. Crit Care Med 20:1127–1133

Adachi S, Kubota M, Matsubara K, Wakazono Y, Hirota H, Kuwakado K, Akiyama Y, Mikawa H (1993) Role of protein kinase C in neutrophil survival enhanced by granulocyte colony-stimulating factor. Exp Hematol 21:1709–1713

Addison IE, Johnson B, Devereux S, Goldstone AH, Linch DC (1989) Granulocyte-macrophage colony-stimulating factor may inhibit neutrophil migration in vivo. Clin Exp Immunol 76:149–153

Aglietta M, Monzeglio C, Pasquino P, Carnino F, Stern AC, Gavosto F (1993) Short-term administration of granulocyte-macrophage colony stimulating factor decreases hematopoietic toxicity of cytostatic drugs. Cancer 72:2970–2973

Aglietta M, Piacibello W, Sanavio F, Stacchini A, Apra F, Schena M, Mossetti C, Carnino F, Caligaris-Cappio F, Gavosto F (1989) Kinetics of human hemopoietic cells after in vivo administration of granulocyte-macrophage colony-stimulating factor. J Clin Invest 83:551–557

Akahane K, Cohen RB, Bickel M, Pluznik DH (1991) IL-1 alpha induces granulocyte-macrophage colony-stimulating factor gene expression in murine B lymphocyte cell lines via mRNA stabilization. J Immunol 146:4190–4196

Akahane K, Pluznik DH (1992) Interleukin-4 inhibits interleukin-1 alpha-induced granulocyte-macrophage colony-stimulating factor gene expression in a murine B-lymphocyte cell line via downregulation of RNA precursor. Blood 79:3188–3195

Akahane K, Pluznik DH (1993) Interferon-gamma destabilizes interleukin-1-induced granulocyte-macrophage colony-stimulating factor mRNA in murine vascular endothelial cells. Exp Hematol 21:878–884

Alam R, Welter JB, Forsythe PA, Lett-Brown MA, Grant JA (1989) Comparative effect of recombinant IL-1, -2, -3, -4, and -6, IFN-gamma, granulocyte-macrophage-colony-stimulating factor, tumor necrosis factor-alpha, and histamine-releasing factors on the secretion of histamine from basophils. J Immunol 142:3431–3435

Allen RC, Stevens PR, Price TH, Chatta GS, Dale DC (1997) In vivo effects of recombinant human granulocyte colony-stimulating factor on neutrophil oxidative functions in normal human volunteers. J Infect Dis 175:1184–1192

Aman MJ, Stockdreher K, Thews A, Kienast K, Aulitzky WE, Farber L, Haus U, Koci B, Huber C, Peschel C (1996) Regulation of immunomodulatory functions by granulocyte-macrophage colony-stimulating factor and granulocyte colony-stimulating factor in vivo. Ann Hematol 73:231–238

Andresen J, Movahhed H (1998) Filgrastim (r-metHuG-CSF) in pneumonia. In: Morstyn G. et al. (eds) Filgrastim in clinical practice, Dekker, New York, pp 429–446

Antman KS, Griffin JD, Elias A, Socinski MA, Ryan L, Cannistra SA, Oette D, Whitley M, Frei E 3d, Schnipper LE (1988) Effect of recombinant human granulocyte-macrophage colony-stimulating factor on chemotherapy-induced myelosuppression [see comments]. N Engl J Med 319:593–598

Aoki Y, Hiromatsu K, Kobayashi N, Hotta T, Saito H, Igarashi H, Niho Y, Yoshikai Y (1995) Protective effect of granulocyte colony-stimulating factor against T-cell-meditated lethal shock triggered by superantigens. Blood 86:1420–1427

Arnaout MA, Wang EA, Clark SC, Sieff CA (1986) Human recombinant granulocyte-macrophage colony-stimulating factor increases cell-to-cell adhesion and surface expression of adhesion-promoting surface glycoproteins on mature granulocytes. J Clin Invest 78:597–601

Asano S, Shirafuji N, Watari K, Matsuda S, Uemura N, Ieki R, Kodo H, Takaku F (1988) Phase I clinical study for recombinant human granulocyte colony-stimulating factor. Behring Inst Mitt 83:222–228

Avalos BR (1996) Molecular analysis of the granulocyte colony-stimulating factor receptor. Blood 88:761–777

Avalos BR, Gasson JC, Hedvat C, Quan SG, Baldwin GC, Weisbart RH, Williams RE, Golde DW, Di Persio JF (1990) Human granulocyte colony-stimulating factor: biologic activities and receptor characterization on hematopoietic cells and small cell lung cancer cell lines. Blood 75:851–857

Azuma I, Ishihara C, Iida J, Yoo YC, Yoshimatsu K, Arikawa J (1992) Stimulation of host-defense mechanism with synthetic adjuvants and recombinant cytokines against viral infection in mice. Adv Exp Med Biol 319:253–263

Bacon K, Gearing A, Camp R (1990) Induction of in vitro human lymphocyte migration by interleukin 3, interleukin 4, and interleukin 6. Cytokine 2:100–105

Badaro R, Nascimento C, Carvalho JS, Badaro F, Russo D, Ho JL, Reed SG, Johnson Jr. WD, Jones TC (1994) Granulocyte-macrophage colony-stimulating factor in combination with pentavalent antimony for the treatment of visceral Leishmaniasis. Eur J Clin Microbiol Infect Dis 13 Suppl 2:S23–S28

Balazovich KJ, Almeida HI, Boxer LA (1991) Recombinant human G-CSF and GM-CSF prime human neutrophils for superoxide production through different signal transduction mechanisms. J Lab Clin Med 118:576–584

Baldwin GC, Chung GY, Kaslander C, Esmail T, Reisfeld RA, Golde DW (1993) Colony-stimulating factor enhancement of myeloid effector cell cytotoxicity towards neuroectodermal tumour cells. Br J Haematol 83:545–553

Baldwin GC, Fuller ND, Roberts RL, Ho DD, Golde DW (1989) Granulocyte- and granulocyte-macrophage colony-stimulating factors enhance neutrophil cytotoxicity toward HIV-infected cells. Blood 74:1673–1677

Baldwin GC, Gasson JC, Quan SG, Fleischmann J, Weisbart R, Oette D, Mitsuyasu RT, Golde DW (1988) Granulocyte-macrophage colony-stimulating factor enhances neutrophil function in acquired immunodeficiency syndrome patients. Proc Natl Acad Sci USA 85:2763–2766

Baram D, Madara P, Culpepper S, Yan L, Malech HL, Danner RL (1996) Human neutrophils regulate TNF concentrations in monocyte cultures and whole blood. Am J Respir Crit Care Med 153

Barber KE, Crosier PS, Watson JD (1987) The differential inhibition of hemopoietic growth factor activity by cytotoxins and interferon-gamma. J Immunol 139:1108–1112

Barsig J, Bundschuh DS, Hartung T, Bauhofer A, Sauer A, Wendel A (1996) Control of fecal peritoneal infection in mice by colony-stimulating factors. J Infect Dis 174:790–799

Baxevanis CN, Dedoussis GVZ, Papadopoulos N,G, Missitzis I, Beroukas C, Stathopoulos GP, Papamichail M (1995) Enhanced human lymphokine- activated killer cell function after brief exposure to granulocyte-macrophage-colony stimulating factor. Cancer 76:1253–1260

Becker S, Warren MK, Haskill S (1987) Colony-stimulating factor-induced monocyte survival and differentiation into macrophages in serum-free cultures. J Immunol 139:3703–3709

Begg SK, Radley JM, Pollard JW, Chisholm OT, Stanley ER, Bertoncello I (1993) Delayed hematopoietic development in osteopetrotic (op/op) mice. J Exp Med 177:237–242

Begley CG, Lopez AF, Nicola NA, Warren DJ, Vadas MA, Sanderson CJ, Metcalf D (1986) Purified colony-stimulating factors enhance the survival of human neutrophils and eosinophils in vitro: a rapid and sensitive microassay for colony-stimulating factors. Blood 68:162–166

Bermudez LE (1994) Potential role of cytokines in disseminated mycobacterial infections. Eur J Clin Microbiol Infect Dis 13 Suppl 2:S29–33

Bermudez LE, Young LS (1990) Recombinant granulocyte-macrophage colony-stimulating factor activates human macrophages to inhibit growth or kill Mycobacterium avium complex. J Leukoc Biol 48:67–73

Bernasconi S, Matteucci C, Sironi M, Conni M, Colotta F, Mosca M, Colombo N, Bonazzi C, Landoni F, Corbetta G, et al (1995) Effects of granulocyte-monocyte colony-stimulating factor (GM-CSF) on expression of adhesion molecules and production of cytokines in blood monocytes and ovarian cancer-associated macrophages. Int J Cancer 60:300–307

Bickel M, Cohen RB, Pluznik DH (1990) Post-transcriptional regulation of granulocyte-macrophage colony-stimulating factor synthesis in murine T cells. J Immunol 145:840–845

Bischoff SC, de Weck AL, Dahinden CA (1990) Interleukin 3 and granulocyte/macrophage-colony-stimulating factor render human basophils responsive to low concentrations of complement component C3a. Proc Natl Acad Sci USA 87:6813–6817

Bober LA, Grace MJ, Pugliese-Sivo C, Rojas-Triana A, Sullivan LM, Narula SK (1995b) The effects of colony stimulating factors on human monocyte cell function. Int J Immunopharmacol 17:385–392

Bober LA, Grace MJ, Pugliese-Sivo C, Rojas-Triana A, Waters T, Sullivan LM, Narula SK (1995a) The effect of GM-CSF and G-CSF on human neutrophil function. Immunopharmacology 29:111–119

Bochner BS, McKelvey AA, Sterbinsky SA, Hildreth JE, Derse CP, Klunk DA, Lichtenstein LM, Schleimer RP (1990) IL-3 augments adhesiveness for endothelium and CD11b expression in human basophils but not neutrophils. J Immunol 145:1832–1837

Bodey GP, Anaissie E, Gutterman J, Vadhan-Raj S (1994) Role of granulocyte-macrophage colony-stimulating factor as adjuvant treatment in neutropenic patients with bacterial and fungal infection. Eur J Clin Microbiol Infect Dis 13 Suppl 2:S18–22

Bokemeyer C, Schmoll H-J (1995) Clinical results obtained with granulocyte-macrophage colony-stimulating factor (GM-CSF) in patients with testicular cancer. In: Nissen NI, Dexter TM (eds) Granulocyte-macrophage colony-stimulating factor (GM-CSF). Current use and future applications. Proceedings from an international symposium, Lucerne, Switzerland, September 9–11 1993. Pennine Press, Cheshire, pp 8–14

Borger P, Kauffman HF, Vijgen JL, Postma DS, Vellenga E (1996) Activation of the cAMP-dependent signaling pathway downregulates the expression of interleukin-3 and granulocyte-macrophage colony-stimulating factor in activated human T lymphocytes. Exp Hematol 24:108–115

Borregaard N, Lollike K, Kjeldsen L, Sengelov H, Bastholm L, Nielsen MH, Bainton DF (1993) Human neutrophil granules and secretory vesicles. Eur J Haematol 51:187–198

Bosse M, Audette M, Ferland C, Pelletier G, Chu HW, Dakhama A, Lavigne S, Boulet LP, Laviolette M (1996) Gene expression of interleukin-2 in purified human peripheral blood eosinophils. Immunology 87:149–154

Brach MA, de Vos S, Gruss HJ, Herrmann F (1992) Prolongation of survival of human polymorphonuclear neutrophils by granulocyte-macrophage colony-stimulating factor is caused by inhibition of programmed cell death. Blood 80:2920–2924

Bronchud MH, Potter MR, Morgenstern G, Blasco MJ, Scarffe JH, Thatcher N, Crowther D, Souza LM, Alton NK, Testa NG (1988) In vitro and in vivo analysis of the effects of recombinant human granulocyte colony-stimulating factor in patients. Br J Cancer 58:64–69

Broudy VC, Kaushansky K, Segal GM, Harlan JM, Adamson JW (1986) Tumor necrosis factor type alpha stimulates human endothelial cells to produce granulocyte/macrophage colony-stimulating factor. Proc Natl Acad Sci USA 83:7467–7471

Broxmeyer HE, Williams DE, Cooper S, Shadduck RK, Gillis S, Waheed A, Urdal DL, Bicknell DC (1987) Comparative effects in vivo of recombinant murine interleukin 3, natural murine colony-stimulating factor-1, and recombinant murine granulocyte-macrophage colony-stimulating factor on myelopoiesis in mice. J Clin Invest 79:721–730

Bundschuh DS, Barsig J, Hartung T, Randow F, Döcke W-D, Volk H-D, Wendel A (1997) Granulocyte-macrophage colony-stimulating factor and IFN-γ restore the systemic TNF-α response to endotoxin in lipopolysaccharide-desensitized mice. J Immun 158:2862–2871

Bussolino F, Wang JM, Defilippi P, Turrini F, Sanavio F, Edgell CJ, Aglietta M, Arese P, Mantovani A (1989) Granulocyte- and granulocyte-macrophage-colony stimulating factors induce human endothelial cells to migrate and proliferate. Nature 337:471–473

Cairo MS (1993) Therapeutic implications of dysregulated colony-stimulating factor expression in neonates [editorial; comment]. Blood 82:2269–2272

Cairo MS, Gillan ER, Buzby JS, van de Ven C, Suen Y (1993) Circulating steel factor (SLF) and G-CSF levels in preterm and term newborn and adult peripheral blood. Am J Pediatr Hematol Oncol 15:311–315

Cairo MS, Mauss D, Kommareddy S, Norris K, van de Ven C, Modanlou H (1990c) Prophylactic or simultaneous administration of recombinant human granulocyte colony stimulating factor in the treatment of group B streptococcal sepsis in neonatal rats. Pediatr Res 27:612–616

Cairo MS, Plunkett JM, Mauss D, Van de ven C (1990b) Seven-day administration of recombinant human granulocyte colony-stimulating factor to newborn rats: modulation of neonatal neutrophilia, myelopoiesis, and group B Streptococcus sepsis. Blood 76:1788–1794

Cairo MS, Plunkett JM, Nguyen A, van de Ven C (1992b) Effect of stem cell factor with and without granulocyte colony-stimulating factor on neonatal hematopoiesis: in vivo induction of newborn myelopoiesis and reduction of mortality during experimental group B streptococcal sepsis. Blood 80:96–101

Cairo MS, Suen Y, Sender L, Gillan ER, Ho W, Plunkett JM, van de Ven C (1992a) Circulating granulocyte colony-stimulating factor (G-CSF) levels after allogeneic and autologous bone marrow transplantation: endogenous G-CSF production correlates with myeloid engraftment. Blood 79:1869–1873

Cairo MS, VandeVen C, Toy C, Mauss D, Sheikh K, Kommareddy S, Modanlou H (1990a) Lymphokines: Enhancement by granulocyte-macrophage and granulocyte colony-stimulating factors of neonatal myeloid kinetics and functional activation of polymorphonuclear leukocytes. Rev Infect Dis 12:S492–S497

Cannistra SA, Griffin JD (1988) Regulation of the production and function of granulocytes and monocytes. Semin Hematol 25:173–188

Cannistra SA, Rambaldi A, Spriggs DR, Herrmann F, Kufe D, Griffin JD (1987) Human granulocyte-macrophage colony-stimulating factor induces expression of the tumor necrosis factor gene by the U937 cell line and by normal human monocytes. J Clin Invest 79:1720–1728

Cannistra SA, Vellenga E, Groshek P, Rambaldi A, Griffin JD (1988) Human granulocyte-monocyte colony-stimulating factor and interleukin 3 stimulate monocyte cytotoxicity through a tumor necrosis factor-dependent mechanism. Blood 71:672–676

Cantrell MA, Anderson D, Cerretti DP, Price V, McKereghan K, Tushinski RJ, Mochizuki DY, Larsen A, Grabstein K, Gillis S, et al (1985) Cloning, sequence, and expression of a human granulocyte/macrophage colony-stimulating factor. Proc Natl Acad Sci USA 82:6250–6254

Caracciolo D, Shirsat N, Wong GG, Lange B, Clark S, Rovera G (1987) Recombinant human macrophage colony-stimulating factor (M-CSF) requires subliminal concentrations of granulocyte/macrophage (GM)-CSF for optimal stimulation of human macrophage colony formation in vitro. J Exp Med 166:1851–1860

Cassatella MA (1995) The production of cytokines by polymorphonuclear neutrophils. Immunol Today 16:21–26

Cebon J, Layton JE, Maher D, Morstyn G (1994) Endogenous haemopoietic growth factors in neutropenia and infection. Br J Haematol 86:265–274

Cenci E, Bartocci A, Puccetti P, Mocci S, Stanley ER, Bistoni F (1991) Macrophage colony-stimulating factor in murine candidiasis: serum and tissue levels during infection and protective effect of exogenous administration. Infect Immun 59:868–872

Chachoua A, Oratz R, Hoogmoed R, Caron D, Peace D, Liebes L, Blum RH, Vilcek J (1994) Monocyte activation following systemic administration of granulocyte-macrophage colony-stimulating factor. J Immunotherapy 15:217–224

Chang JM, Metcalf D, Gonda TJ, Johnson GR (1989) Long-term exposure to retrovirally expressed granulocyte-colony-stimulating factor induces a nonneoplastic granulocytic and progenitor cell hyperplasia without tissue damage in mice. J Clin Invest 84:1488–1496

Chantry D, Turner M, Brennan F, Kingsbury A, Feldmann M (1990) Granulocyte-macrophage colony stimulating factor induces both HLA-DR expression and cytokine production by human monocytes. Cytokine 2:60–67

Cheers C, Haigh AM, Kelso A, Metcalf D, Stanley ER, Young AM (1988) Production of colony-stimulating factors (CSFs) during infection: separate determinations of macrophage-, granulocyte-, granulocyte-macrophage-, and multi-CSFs. Infect Immun 56:247–251

Chen BD, Mueller M, Chou TH (1988) Role of granulocyte/macrophage colony-stimulating factor in the regulation of murine alveolar macrophage proliferation and differentiation. J Immunol 141:139–144

Cheung DL, Hamilton JA (1992) Regulation of human monocyte DNA synthesis by colony-stimulating factors, cytokines, and cyclic adenosine monophosphate. Blood 79:1972–1981

Chikkappa G, Broxmeyer HE, Cooper S, Williams DE, Hangoc G, Greenberg ML, Waheed A, Shadduck RK (1989) Effect in vivo of multiple injections of purified murine and recombinant human macrophage colony-stimulating factor to mice. Cancer Res 49:3558–3561

Churchill L, Friedman B, Schleimer RP, Proud D (1992) Production of granulocyte-macrophage colony-stimulating factor by cultured human tracheal epithelial cells. Immunology 75:189–195

Cicco NA, Lindemann A, Content J, Vandenbussche P, Lubbert M, Gauss J, Mertelsmann R, Herrmann F (1990) Inducible production of interleukin-6 by human polymorphonuclear neutrophils: role of granulocyte-macrophage colony-stimulating factor and tumor necrosis factor-alpha. Blood 75:2049–2052

Clark SC (1988) Biological activities of human granulocyte-macrophage colony-stimulating factor. Int J Cell Cloning he 6:365–377

Clark SC, Kamen R (1987) The human hematopoietic colony-stimulating factors. Science 236:1229–1237

Cluitmans FH, Esendam BH, Landegent JE, Willemze R, Falkenburg JH (1993) Regulatory effects of T cell lymphokines on cytokine gene expression in monocytes. Lymphokine Cytokine Res 12:457–464

Coffey RG (1989) Mechanism of GM-CSF stimulation of neutrophils. Immunol Res 8:236–48

Coffey RG, Diaz S, Coffey CS (1993) Effects of inhibition of lipoxygenase and guanylate cyclase on human neutrophil responses to formyl peptide and granulocyte-macrophage colony-stimulating factor. J Leukoc Biol 54:89–96

Cohen AM, Hines DK, Korach ES, Ratzkin BJ (1988) In vivo activation of neutrophil function in hamsters by recombinant human granulocyte colony-stimulating factor. Infect Immun 56:2861–2865

Coleman DL, Chodakewitz JA, Bartiss AH, Mellors JW (1988) Granulocyte-macrophage colony-stimulating factor enhances selective effector functions of tissue-derived macrophages. Blood 72:573–578

Collins HL, Bancroft GJ (1992) Cytokine enhancement of complement-dependent phagocytosis by macrophages: synergy of tumor necrosis factor-alpha and granulocyte-macrophage colony-stimulating factor for phagocytosis of Cryptococcus neoformans. Eur J Immunol 22:1447–1454

Colotta F, Re F, Polentarutti N, Sozzani S, Mantovani A (1992) Modulation of granulocyte survival and programmed cell death by cytokines and bacterial products. Blood 80:2012–2020

Conti P, Reale M, Barbacane RC, Panara MR, Bongrazio M, Fiore S, Mier JW, Dempsey RA (1992) Granulocyte-macrophage colony stimulating factor potentiates human polymorphonuclear leukocyte aggregation responses to formyl-methionyl-leucyl-phenylalanine. Immunol Lett 32:71–79

Crapper RM, Vairo G, Hamilton JA, Clark-Lewis I, Schrader JW (1985) Stimulation of bone marrow-derived and peritoneal macrophages by a T lymphocyte-derived hemopoietic growth factor, persisting cell-stimulating factor. Blood 66:859–865

Cuturi MC, Anegon I, Sherman F, Loudon R, Clark SC, Perussia B, Trinchieri G (1989) Production of hematopoietic colony-stimulating factors by human natural killer cells. J Exp Med 169:569–583

Dahinden CA, Zingg J, Maly FE, de Weck AL (1988) Leukotriene production in human neutrophils primed by recombinant human granulocyte/macrophage colony-stimulating factor and stimulated with the complement component C5A and FMLP as second signals. J Exp Med 167:1281–1295

Dale DC (1994) Potential role of colony-stimulating factors in the prevention and treatment of infectious diseases. Clin Infect Dis 18 Suppl 2:S180–188

Dale DC, Lau S, Nash R, Boone T, Osborne W (1992) Effect of endotoxin on serum granulocyte and granulocyte-macrophage colony-stimulating factor levels in dogs [see comments]. J Infect Dis 165:689–694

Dale DC, Liles WC, Summer WR, Nelson S (1995) Review: granulocyte colony-stimulating factor-role and relationships in infectious diseases. J Infect Dis 172:1061–1075

D'Alesandro MM, Gruber DF, O'Halloran KP, MacVittie TJ (1991) In vitro modulation of canine polymorphonuclear leukocyte function by granulocyte-macrophage colony stimulating factor. Biotherapy 3:233–9

Danis VA, Franic GM, Rathjen DA, Brooks PM (1991) Effects of granulocyte-macrophage colony-stimulating factor (GM-CSF), IL-2, interferon-gamma (IFN-gamma), tumour necrosis factor-alpha (TNF-alpha) and IL-6 on the production of immunoreactive IL-1 and TNF-alpha by human monocytes. Clin Exp Immunol 85:143–150

Daschner FD, Grundmann H, Anding K, Lemmen S (1995) Combined effect of human neutrophils, ceftazidime and granulocyte colony-stimulating factor on killing of Escherichia coli. Eur J Clin Microbiol Infect Dis 14:536–539

de Andres B, Cardaba B, del Pozo V, Martin-Orozco E, Gallardo S, Tramon P, Palomino P, Lahoz C (1994) Modulation of the Fc gamma RII and Fc gamma RIII induced by GM-CSF, IFN-gamma and IL-4 on murine eosinophils. Immunology 83:155–160

de Haas M, Kerst JM, van der Schoot CE, Calafat J, Hack CE, Nuijens JH, Roos D, van Oers RH, von dem Borne AE (1994) Granulocyte colony-stimulating factor administration to healthy volunteers: analysis of the immediate activating effects on circulating neutrophils. Blood 84:3885–3894

de Wit H, Dokter WH, Esselink MT, Halie MR, Vellenga E (1993) Interferon-gamma enhances the LPS-induced G-CSF gene expression in human adherent monocytes, which is regulated at transcriptional and posttranscriptional levels. Exp Hematol 21:785–790

De Witte T, Vreugdenhil A, Schattenberg A (1995) The role of recombinant granulocyte-macrophage colony-stimulating factor (GM-CSF) after T-cell depleted allogeneic bone marrow transplantation. In: Nissen NI, Dexter TM (eds) Granulocyte-macrophage colony-stimulating factor (GM-CSF). Current use and future applications. Procceedings from an international symposium, Lucerne, Switzerland, September 9–11 1993. Pennine Press, Cheshire, pp 30–34

Demetri GD, Griffin JD (1991) Granulocyte colony-stimulating factor and its receptor. Blood 78:2791–2808

Demetri GD, Spertini O, Pratt ES, Kim E, Elias A, Antman K, Tedder TF, Griffin JD (1990) GM-CSF and G-CSF Have Different Effects on Expression of the Neutrophil Adhesion Receptors LAM-1 & CD11b. Blood, Supplement 76:704a

Demetri GD, Zenzie BW, Rheinwald JG, Griffin JD (1989) Expression of colony-stimulating factor genes by normal human mesothelial cells and human malignant mesothelioma cells lines in vitro. Blood 74:940–946

Denis M (1991a) Tumor necrosis factor and granulocyte macrophage-colony stimulating factor stimulate human macrophages to restrict growth of virulent Mycobacterium avium and to kill avirulent M. avium: killing effector mechanism depends on the generation of reactive nitrogen intermediates. J Leukoc Biol 49:380–387

Denis M (1991b) Growth of Mycobacterium avium in human monocytes: identification of cytokines which reduce and enhance intracellular microbial growth. Eur J Immunol 21:391–395

Denis M, Gregg EO (1991) Identification of cytokines which enhance (CSF-1, IL-3) or restrict (IFN-gamma) growth of intramacrophage Listeria monocytogenes. Immunol Lett 27:237–242

Denzlinger C, Holler E, Reisbach G, Hiller E, Wilmanns W (1994) Granulocyte colony-stimulating factor inhibits the endogenous leukotriene production in tumour patients. Br J Haematol 86:881–882

Denzlinger C, Tetzloff W, Gerhartz HH, Pokorny R, Sagebiel S, Haberl C, Wilmanns W (1993) Differential activation of the endogenous leukotriene biosynthesis by two different preparations of granulocyte-macrophage colony-stimulating factor in healthy volunteers. Blood 81:2007–2013

Dessein AJ, Vadas MA, Nicola NA, Metcalf D, David JR (1982) Enhancement of human blood eosionophil cytotoxicity by semi-purified eosinophil colony-stimulating factor (s). J Exp Med 156:90–103

Devereux S, Bull HA, Campos-Costa D, Saib R, Linch DC (1989) Granulocyte macrophage colony stimulating factor induced changes in cellular adhesion molecule expression and adhesion to endothelium: in-vitro and in-vivo studies in man. Br J Haematol 71:323–330

Dexter TM, Heyworth CM (1994) Growth factors and the molecular control of haematopoiesis. Eur J Clin Microbiol Infect Dis 13 Suppl 2:S3–8

Di Persio J, Billing P, Kaufman S, Eghtesady P, Williams RE, Gasson JC (1988a) Characterization of the human granulocyte-macrophage colony-stimulating factor receptor. J Biol Chem 263:1834–1841

Di Persio JF (1990) Colony-stimulating factors: enhancement of effector cell function. Cancer Surv 9:81–113

Di Persio JF, Billing P, Williams R, Gasson JC (1988c) Human granulocyte-macrophage colony-stimulating factor and other cytokines prime human neutrophils for enhanced arachidonic acid release and leukotriene B4 synthesis. J Immunol 140:4315–4322

Di Persio JF, Naccache PH, Borgeat P, Gasson JC, Nguyen MH, McColl SR (1988b) Characterization of the priming effects of human granulocyte-macrophage colony-stimulating factor on human neutrophil leukotriene synthesis. Prostaglandins 36:673–691

DiPersio JF, Abboud CN (1992) Activation of neutrophils by granulocyte-macrophage colony-stimulating factor. In: Coffey RG (ed) Granulocyte responses to cytokines. Basic and clinical research. Marcel Dekker, Inc., New York; Basel; Hong Kong, pp 457–84

Djeu JY, Widen R, Blanchard DK (1989) Susceptibility of monocytes to lymphokine-activated killer cell lysis: effect of granulocyte-macrophage colony-stimulating factor and interleukin-3. Blood 73:1264–1271

Dokter WH, Dijkstra AJ, Koopmans SB, Mulder AB, Stulp BK, Halie MR, Keck W, Vellenga E (1994) G(AnH)MTetra, a naturally occurring 1,6-anhydro muramyl dipeptide, induces granulocyte colony-stimulating factor expression in human monocytes: a molecular analysis. Infect Immun 62:2953–2957

Donahue RE, Emerson SG, Wang EA, Wong GG, Clark SC, Nathan DG (1985) Demonstration of burst-promoting activity of recombinant human GM-CSF on circulating erythroid progenitors using an assay involving the delayed addition of erythropoietin. Blood 66:1479–1481

Donahue RE, Wang EA, Stone DK, Kamen R, Wong GG, Sehgal PK, Nathan DG, Clark SC (1986) Stimulation of haematopoiesis in primates by continuous infusion of recombinant human GM-CSF. Nature 321:872–875

Dong F, Dale DC, Bonilla MA, Freedman M, Fasth A, Neijens HJ, Palmblad J, Briars GL, Carlsson G, Veerman AJ, Welte K, Lowenberg B, Touw IP (1997) Mutations in the granulocyte colony-stimulating factor receptor gene in patients with severe congenital neutropenia. Leukemia 11:120–125

Dong F, Hoefsloot LH, Schelen AM, Broeders CA, Meijer Y, Veerman AJ, Touw IP, Lowenberg B (1994) Identification of a nonsense mutation in the granulocyte-colony-stimulating factor receptor in severe congenital neutropenia. Proc Natl Acad Sci USA 91:4480–4484

Dranoff G, Jaffee E, Lazenby A, Golumbek P, Levitsky H, Brose K, Jackson V, Hamada H, Pardoll D, Mulligan RC (1993) Vaccination with irradiated tumor cells engineered to secrete murine granulocyte-macrophage colony-stimulating factor stimulates potent, specific, and long-lasting anti-tumor immunity. Proc Natl Acad Sci USA 90:3539–3543

Dreger P, Grelle K, Eckstein V, Suttorp M, Muller-Ruchholtz W, Loffler H, Schmitz N (1993) Granulocyte-colony-stimulating factor induces increased serum levels of soluble interleukin 2 receptors preceding engraftment in autologous bone marrow transplantation. Br J Haematol 83:7–13

Duhrsen U, Villeval JL, Boyd J, Kannourakis G, Morstyn G, Metcalf D (1988) Effects of recombinant human granulocyte colony-stimulating factor on hematopoietic progenitor cells in cancer patients. Blood 72:2074–2081

Dunne JR, Dunkin BJ, Nelson S, White JC (1996) Effects of granulocyte colony stimulating factor in an nonneutropenic rodent model of Escherichia coli peritonitis. J Surg Res 61:348–354

Eichacker PQ, Waisman Y, Natanson C, Farese A, Hoffman WD, Banks SM, MacVittie TJ (1994) Cardiopulmonary effects of granulocyte colony-stimulating factor in a canine model of bacterial sepsis. J Appl Physiol 77:2366–2373

Elliott MJ, Vadas MA, Eglinton JM, Park LS, To LB, Cleland LG, Clark SC, Lopez AF (1989) Recombinant human interleukin-3 and granulocyte-macrophage colony-stimulating factor show common biological effects and binding characteristics on human monocytes. Blood 74:2349–2359

Elsasser D, Valerius T, Repp R, Weiner GJ, Deo Y, Kalden JR, van de Winkel JG, Stevenson GT, Glennie MJ, Gramatzki M (1996) HLA class II as potential target antigen on malignant B cells for therapy with bispecific antibodies in combination with granulocyte colony-stimulating factor. Blood 87:3803–3812

Endo S, Inada K, Inoue Y, Yamada Y, Takakuwa T, Kasai T, Nakae H, Kuwata Y, Hoshi S, Yashida M (1994) Evaluation of recombinant human granulocyte colony-stimulating factor (rhG-CSF) therapy in granulopoetic patients complicated with sepsis. Curr Med Res Opin 13:233–241

Endo Y, Kikuchi T, Takeda Y, Nitta Y, Rikiishi H, Kumagai K (1992) GM-CSF and G-CSF stimulate the synthesis of histamine and putrescine in the hematopoietic organs in vivo. Immunol Lett 33:9–13

English D, Broxmeyer HE, Gabig TG, Akard LP, Williams DE, Hoffman R (1988) Temporal adaptation of neutrophil oxidative responsiveness to n-formylmethionyl-leucyl-phenylalanine. Acceleration by granulocyte-macrophage colony stimulating factor. J Immunol 141:2400–2406

Ernst TJ, Ritchie AR, Demetri GD, Griffin JD (1989) Regulation of granulocyte- and monocyte-colony stimulating factor mRNA levels in human blood monocytes is mediated primarily at a post-transcriptional level. J Biol Chem 264:5700–5703

Esser R, Glienke W, von Briesen H, Rubsamen-Waigmann H, Andreesen R (1996) Differential regulation of proinflammatory and hematopoietic cytokines in human macrophages after infection with human immunodeficiency virus. Blood 88:3474–3481

Evans R, Kamdar SJ, Duffy TM, Fuller J (1992) Synergistic interaction of bacterial lipopolysaccharide and the monocyte-macrophage colony-stimulating factor: potential quantitative and qualitative changes in macrophage-produced cytokine bioactivity. J Leukoc Biol 51:93–96

Fabian I, Kletter Y, Mor S, Geller-Bernstein C, Ben-Yaakov M, Volovitz B, Golde DW (1992) Activation of human eosinophil and neutrophil functions by haematopoietic growth factors: comparisons of IL-1, IL-3, IL-5 and GM-CSF. Br J Haematol 80:137–143

Fan K, Ruan Q, Sensenbrenner L, Chen BD (1992) Up-regulation of granulocyte-macrophage colony-stimulating factor (GM-CSF) receptors in murine peritoneal exudate macrophages by both GM-CSF and IL-3. J Immunol 149:96–102

Fattorini L, Xiao Y, Li B, Santoro C, Ippoliti F, Orefici G (1994) Induction of IL-1 beta, IL-6, TNF-alpha, GM-CSF and G-CSF in human macrophages by smooth transparent and smooth opaque colonial variants of Mycobacterium avium. J Med Microbiol 40:129–133

Feder LS, Laskin DL (1994) Regulation of hepatic endothelial cell and macrophage proliferation and nitric oxide production by GM-CSF, M-CSF, and IL-1 beta following acute endotoxemia. J Leukoc Biol 55:507–513

Fernandez MC, Walters J, Marucha P (1996) Transcriptional and post-transcriptional regulation of GM-CSF-induced IL-1 beta gene expression in PMN. J Leukoc Biol 59:598–603

Fibbe WE, van Damme J, Billiau A, Goselink HM, Voogt PJ, van Eeden G, Ralph P, Altrock BW, Falkenburg JH (1988) Interleukin 1 induces human marrow stromal cells in long-term culture to produce granulocyte colony-stimulating factor and macrophage colony-stimulating factor. Blood 71:430–435

Filonzi EL, Zoellner H, Stanton H, Hamilton JA (1993) Cytokine regulation of granulocyte-macrophage colony stimulating factor and macrophage colony-stimulating factor production in human arterial smooth muscle cells. Atherosclerosis 99:241–252

Fink MP, O'Sullivan BP, Menconi MJ, Wollert SP, Wang H, Youssef ME, Bellelsle JM (1993) Effect of granulocyte colony-stimulating factor on systemic and pulmonary responses to endotoxin in pigs. J Trauma 34:571–7; discussion 577–8

Fischer HG, Frosch S, Reske K, Reske-Kunz AB (1988) Granulocyte-macrophage colony-stimulating factor activates macrophages derived from bone marrow cultures to synthesis of MHC class II molecules and to augmented antigen presentation function. J Immunol 141:3882–3888

Fitzpatrick DR, Kelso A (1995) Dissociated expression of granulocyte-macrophage CSF and IL-3 in short-term T cell clones from normal mice. J Immunol 155:5140–5150

Fleischmann J, Golde DW, Weisbart RH, Gasson JC (1986) Granulocyte-macrophage colony-stimulating factor enhances phagocytosis of bacteria by human neutrophils. Blood 68:708–711

Foster PF, Mital D, Sankary HN, McChesney LP, Marcon J, Koukoulis G, Kociss K, Leurgans S, Whiting JF, Williams JW (1995) The use of granulocyte colony-stimulating factor after liver transplantation. Transplantation 59:1557–1563

Frampton JE, Yarker YE, Goa KL (1995) Lenograstim. A review of its pharmacological properties and therapeutic efficacy in neutropenia and related clinical settings. Drugs 49:767–793

Frank MO, Mandell GL (1995) Adjuncts to antibacterial therapy. Infect Dis Clin North Am 9:769–781

Freeman BD, Correa R, Karzai W, Natanson C, Patterson M, Banks S, Fitz Y, Danner RL, Wilson L, Eichacker PQ (1996) Controlled Trials of rG-CSF and CD11b-directed MAb during Hyperoxia and E. Coli Pneumonia in Rats. J Appl Physiol 80:2066–2076

Frenck RW, Sarman G, Harper TE, Buescher ES (1990) The ability of recombinant murine granulocyte-macrophage colony-stimulating factor to protect neonatal rats from septic death due to Staphylococcus aureus. J Infect Dis 162:109–114

Frendl G (1992) Interleukin 3: from colony-stimulating factor to pluripotent immunoregulatory cytokine. Int J Immunopharmacol 14:421–430

Frendl G, Beller DI (1990) Regulation of macrophage activation by IL-3. I. IL-3 functions as a macrophage-activating factor with unique properties, inducing IA and lymphocyte function-associated antigen-1 but not cytotoxicity. J Immunol 144:3392–3399

Frendl G, Fenton MJ, Beller DI (1990) Regulation of macrophage activation by IL-3. II. IL-3 and lipopolysaccharide act synergistically in the regulation of IL-1 expression. J Immunol 144:3400–3410

Freund M, Kleine HD (1992) The role of GM-CSF in infection. Infection 20 Suppl 2:S84–92

Frumkin LR (1997) Role of granulocyte colony-stimulating factor and granulocyte-macrophage colony-stimulating factor in the treatment of patients with HIV infection. Curr Opin Hematol 4:200–206

Fukunaga R, Ishizaka-Ikeda E, Nagata S (1993) Growth and differentiation signals mediated by different regions in the cytoplasmic domain of granulocyte colony-stimulating factor receptor. Cell 74:1079–1087

Furman WL, Lumm W, Arnold B, Pratt CB, Ihle JN (1992) Recombinant human TNF-alpha stimulates the secretion of granulocyte colony-stimulating factor in vivo. Leukemia 6:319–322

Gabrilove J (1992) The development of granulocyte colony-stimulating factor in its various clinical applications [editorial]. Blood 80:1382–1385

Gabrilove JL, Jakubowski A (1990) Hematopoietic growth factors: biology and clinical application. J Natl Cancer Inst Monogr 10:73–77

Gabrilove JL, Jakubowski A, Fain K, Grous J, Scher H, Sternberg C, Yagoda A, Clarkson B, Bonilla MA, Oettgen HF (1988) Phase I study of granulocyte colony-stimulating factor in patients with transitional cell carcinoma of the urothelium. J Clin Invest 82:1454–1461

Galy AH, Spits H (1991) IL-1, IL-4, and IFN-gamma differentially regulate cytokine production and cell surface molecule expression in cultured human thymic epithelial cells. J Immunol 147:3823–3830

Galy AH, Spits H, Hamilton JA (1993) Regulation of M-CSF production by cultured human thymic epithelial cells. Lymphokine Cytokine Res 12:265–270

Gamble JR, Elliott MJ, Jaipargas E, Lopez AF, Vadas MA (1989) Regulation of human monocyte adherence by granulocyte-macrophage colony-stimulating factor. Proc Natl Acad Sci USA 86:7169–7173

Ganser A, Seipelt G, Verbeek W, Ottmann OG, Maurer A, Kolbe K, Hess U, Elsner S, Reutzel R, Wormann B (1994) Effect of combination therapy with all-trans-retinoic acid and recombinant human granulocyte colony-stimulating factor in patients with myelodysplastic syndromes. Leukemia 8:369–375

Garnick MB, O´ Reilly RJ (1989) Clinical promise of new hematopoietic growth factors: M-CSF, IL-3, IL-6. Hematol Oncol Clin North Am 3:495–509

Gasson JC (1991) Molecular physiology of granulocyte-macrophage colony-stimulating factor. Blood 77:1131–1145

Gasson JC, Weisbart RH, Kaufman SE, Clark SC, Hewick RM, Wong GG, Golde DW (1984) Purified human granulocyte-macrophage colony-stimulating factor: direct action on neutrophils. Science 226:1339–1342

Gatti E, Scagliotti GV, Ferrari G, Mutti L, Pozzi E, Stacchini A, Fubini L, Aglietta M (1995) Blood cell redistribution in the lung after administration of recombinant human granulocyte-macrophage colony-stimulating factor. Eur Respir J 8:1566–1571

Gazzola MV, Collins NH, Tafuri A, Keever CA (1992) Recombinant interleukin 3 induces interleukin 2 receptor expression on early myeloid cells in normal human bone marrow. Exp Hematol 20:201–208

Gearing DP, King JA, Gough NM, Nicola NA (1989) Expression cloninghe of a receptor for human granulocyte-macrophage colony-stimulating factor. EMBO J 8:3667–3676

Gennari R, Alexander JW, Gianotti L, Eaves-Pyles T, Hartmann S (1994) Granulocyte macrophage colony-stimulating factor improves survival in two models of gut-derived sepsis by improving gut barrier function and modulating bacterial clearance. Ann Surg 220:68–76

Gerhartz HH (1994) Einsatz von rhGM-CSFbei malignen non-Hodgkin-Lymphome. In: Hiddemann W (ed) Klinische Anwendung von rh GM-CSF: Neue Möglichkeiten in der Behandlung von Tumorpatienten. Satelliten-Symposium, 11.03.1994. Hamburg,

Gerhartz HH, Engelhard M, Meusers P, Brittinger G, Wilmanns W, Schlimok G (1995) GM-CSD as adjunct to induction treatment of aggressive Non-Hodgkin´s lymphomas. In: Nissen NI, Dexter TM (eds) Granulocyte-macrophage colony-stimulating factor (GM-CSF). Current use and future applications. Procceedings from an international symposium, Lucerne, Switzerland, September 9–11 1993. Pennine Press, Cheshire, pp 14–20

Gerrard TL, Dyer DR, Mostowski HS (1990) IL-4 and granulocyte-macrophage colony-stimulating factor selectively increase HLA-DR and HLA-DP antigens but not HLA-DQ antigens on human monocytes. J Immunol 144:4670–4674

Gillan E, Plunkett M, Cairo MS (1993) Colony-stimulating factors in the modulation of sepsis. New Horiz 1:96–109

Gillan ER, Christensen RD, Suen Y, Ellis R, van de Ven C, Cairo MS (1994) A randomized, placebo-controlled trial of recombinant human granulocyte colony-stimulating factor administration in newborn infants with presumed sepsis: significant induction of peripheral and bone marrow neutrophilia. Blood 84:1427–1433

Gilmore GL, De Pasquale DK, Fischer BC, Shadduck RK (1995) Enhancement of monocytopoiesis by granulocyte colony-stimulating factor: evidence for secondary cytokine effects in vivo [see comments]. Exp Hematol 23:1319–1323

Givner LB, Nagaraj SK (1993) Hyperimmune human IgG or recombinant human granulocyte-macrophage colony-stimulating factor as adjunctive therapy for group B streptococcal sepsis in newborn rats. J Pediatr 122:774–779

Glaspy JA, Golde DW (1992) Granulocyte colony-stimulating factor (G-CSF): preclinical and clinical studies. Semin Oncol 19:386–394

Gliniak BC, Rohrschneider LR (1990) Expression of the M-CSF receptor is controlled posttranscriptinally by the dominant actions of GM-CSF or multi-CSF. Cell 63:1073–1083

Görgen I, Hartung T, Leist M, Niehorster M, Tiegs G, Uhlig S, Weitzel F, Wendel A (1992) Granulocyte colony-stimulating factor treatment protects rodents against lipopolysaccharide-induced toxicity via suppression of systemic tumor necrosis factor-alpha. J Immunol 149:918–924

Gorin N-G (1995) Granulocyte-macrophage colony-stimulating factor as as adjunct to autologus bone marrow transplantation in haematology: considerations on increased safety in the transplant ward. In: Nissen NI, Dexter TM (eds) Granulocyte-macrophage colony-stimulating factor (GM-CSF). Current use and future applications. Procceedings from an international symposium, Lucerne, Switzerland, September 9–11 1993. Pennine Press, Cheshire, pp 38–46

Gosselin EJ, Wardwell K, Rigby WF, Guyre PM (1993) Induction of MHC class II on human polymorphonuclear neutrophils by granulocyte/macrophage colony-stimulating factor, IFN-gamma, and IL-3. J Immunol 151:1482–1490

Gough NM, Gough J, Metcalf D, Kelso A, Grail D, Nicola NA, Burgess AW, Dunn AR (1984) Molecular cloning of cDNA encoding a murine haematopoietic growth regulator, granulocyte-macrophage colony stimulating factor. Nature 309:763–767

Goya T, Torisu M, Doi F, Yoshida T (1993) Effects of granulocyte colony stimulating factor and monobactam antibiotics (Aztreonam) on neutrophil functions in sepsis. Clin Immunol Immunopathol 69:278–284

Grabstein K, Mochizuki D, Kronheim S, Price V, Cosman D, Urdal D, Gillis S, Conlon P (1986b) Regulation of antibody production in vitro by granulocyte-macrophage colony stimulating factor. J Mol Cell Immunol 2:199–207

Grabstein KH, Urdal DL, Tushinski RJ, Mochizuki DY, Price VL, Cantrell MA, Gillis S, Conlon PJ (1986a) Induction of macrophage tumoricidal activity by granulocyte-macrophage colony-stimulating factor. Science 232:506–508

Granowitz EV, Porat R, Orencole SF, Callahan MV, Lynch EA, Wolff SM, Dinarello CA (1992) Granulocyte-macrophage colony-stimulating factor synthesis during experimental endotoxemia in humans [letter; comment]. J Infect Dis 166:1204–1205

Grau GE, Kindler V, Piguet PF, Lambert PH, Vassalli P (1988) Prevention of experimental cerebral malaria by anticytokine antibodies. Interleukin 3 and granulocyte macrophage colony-stimulating factor are intermediates in increased tumor necrosis factor production and macrophage accumulation. J Exp Med 168:1499–1504

Griffin JD, Spertini O, Ernst TJ, Belvin MP, Levine HB, Kanakura Y, Tedder TF (1990) Granulocyte-macrophage colony-stimulating factor and other cytokines regulate surface expression of the leukocyte adhesion molecule-1 on human neutrophils, monocytes, and their precursors. J Immunol 145:576–584

Grigg A, Bardy P, Vecchi L, Szer J (1996) G-CSF stimulated donor granulocyte collections for prophylaxis and therapy of neutropenic sepsis. Aust NZ J Med 26:813–8☒

Gruber MF, Gerrard TL (1992) Production of macrophage colony-stimulating factor (M-CSF) by human monocytes is differentially regulated by GM-CSF, TNF alpha, and IFN-gamma. Cell Immunol 142:361–369

Gruber MF, Webb DS, Gerrard TL (1992) Stimulation of human monocytes via CD45, CD44, and LFA-3 triggers macrophage-colony-stimulating factor production. Synergism with lipopolysaccharide and IL-1 beta. J Immunol 148:1113–1118

Gruber MF, Weih KA, Boone EJ, Smith PD, Clouse KA (1995) Endogenous macrophage CSF production is associated with viral replication in HIV-1-infected human monocyte-derived macrophages. J Immunol 154:5528–5535

Gualtieri RJ, Liang CM, Shadduck RK, Waheed A, Banks J (1987) Identification of the hematopoietic growth factors elaborated by bone marrow stromal cells using antibody neutralization analysis. Exp Hematol 15:883–889

Gulati SC (1995) Side effects of granulocyte-macrophage colony-stimulating factor. In: Nissen NI, Dexter TM (eds) Granulocyte-macrophage colony-stimulating factor (GM-CSF). Current use and future applications. Procceedings from an international symposium, Lucerne, Switzerland, September 9–11 1993. Pennine Press, Cheshire, pp 55–64

Haak-Frendscho M, Arai N, Arai K, Baeza ML, Finn A, Kaplan AP (1988) Human recombinant granulocyte-macrophage colony-stimulating factor and interleukin 3 cause basophil histamine release. J Clin Invest 82:17–20

Hallsworth MP, Giembycz MA, Barnes PJ, Lee TH (1996) Cyclic AMP-elevating agents prolong or inhibit eosinophil survival depending on prior exposure to GM-CSF. Br J Pharmacol 117:79–86

Hamblin AS (1988) Lymphokines and interleukins. Immunol Suppl 1:39–44

Hamilton JA (1994) Coordinate and noncoordinate colony stimulating factor formation by human monocytes. J Leukoc Biol 55:355–361

Hamilton JA, Piccoli DS, Cebon J, Layton JE, Rathanaswani P, McColl SR, Leizer T (1992a) Cytokine regulation of colony-stimulating factor (CSF) production in cultured human synovial fibroblasts. II. Similarities and differences in the control of interleukin-1 induction of granulocyte-macrophage CSF and granulocyte-CSF production. Blood 79:1413–1419

Hamilton JA, Vairo G, Knight KR, Cocks BG (1991) Activation and proliferation signals in murine macrophages. Biochemical signals controlling the regulation of macrophage urokinase-type plasminogen activator activity by colony-stimulating factors and other agents. Blood 77:616–627

Hamilton JA, Whitty GA, Royston AK, Cebon J, Layton JE (1992b) Interleukin-4 suppresses granulocyte colony-stimulating factor and granulocyte-macrophage colony-stimulating factor levels in stimulated human monocytes. Immunology 76:566–571

Hamilton JA, Whitty GA, Stanton H, Wojta J, Gallichio M, McGrath K, Ianches G (1993) Macrophage colony-stimulating factor and granulocyte-macrophage colony-stimulating factor stimulate the synthesis of plasminogen-activator inhibitors by human monocytes. Blood 82:3616–3621

Hammer SM, Gillis JM, Groopman JE, Rose RM (1986) In vitro modification of human immunodeficiency virus infection by granulocyte-macrophage colony-stimulating factor and gamma interferon. Proc Natl Acad Sci USA 83:8734–8738

Hammond WP, Boone TC, Donahue RE, Souza LM, Dale DC (1990) A comparison of treatment of canine cyclic hematopoiesis with recombinant human granulocyte-macrophage colony-stimulating factor (GM-CSF), G-CSF interleukin-3, and canine G-CSF. Blood 76:523–532

Hammond WP, Csiba E, Canin A, Hockman H, Souza LM, Layton JE, Dale DC (1991) Chronic neutropenia. A new canine model induced by human granulocyte colony-stimulating factor. J Clin Invest 87:704–710

Hamood M, Bluche PF, De Vroey C, Corazza F, Bujan W, Fondu P (1994) Effects of recombinant human granulocyte-colony stimulating factor on neutropenic mice infected with Candida albicans: acceleration of recovery from neutropenia and potentiation of anti-C. albicans resistance. Mycoses 37:93–99

Hancock WW, Pleau ME, Kobzik L (1988) Recombinant granulocyte-macrophage colony-stimulating factor down-regulates expression of IL-2 receptor on human mononuclear phagocytes by induction of prostaglandin E. J Immunol 140:3021–3025

Handman E, Burgess AW (1979) Stimulation by granulocyte-macrophage colony-stimulating factor of Leishmania tropica killing by macrophages. J Immunol 122:1134–1137

Hansen PB, Kjaersgaard E, Johnsen HE, Gram J, Pedersen M, Nikolajsen K, Hansen NE (1993) Different membrane expression of CD11b and CD14 on blood neutrophils following in vivo administration of myeloid growth factors [published erratum appears in Br J Haematol 1994 Jan; 86(1):240]. Br J Haematol 85:50–56

Hart PH, Whitty GA, Burgess DR, Hamilton JA (1990) Regulation by interleukin-3 of human monocyte pro-inflammatory mediators. Similarities with granulocyte-macrophage colony-stimulating factor. Immunology 71:76–82

Hart PH, Whitty GA, Piccoli DS, Hamilton JA (1988) Synergistic activation of human monocytes by granulocyte-macrophage colony-stimulating factor and IFN-gamma. Increased TNF-alpha but not IL-1 activity. J Immunol 141:1516–1521

Hartung T, Docke WD, Gantner F, Krieger G, Sauer A, Stevens P, Volk HD, Wendel A (1995a) Effect of granulocyte colony-stimulating factor treatment on ex vivo blood cytokine response in human volunteers. Blood 85:2482–2489

Hartung T, Volk H-D,Wendel,A (1995b) G-CSF – an anti-inflammatory cytokine. J Endo Res 2:195–201

Hartung T, Wendel A (1998) Immunomodulatory properties of Filgrastim (r-met-HuG-CSF) in preclinical models. In: Morstyn et al. (eds) Filgrastim in clinical practice, Dekker, New York, pp 397–427

Hashimoto S, Gon Y, Hayashi S, Tomita Y, Yodoi J, Horie T (1997) TNF-alpha regulates GM-CSF-, IL-3- or M-CSF-induced Fc epsilon RII/CD23 gene expression and soluble Fc epsilon RII release by human monocytes. Scand J Immunol 45:255–260

Hast R, Bernell P, Kimby E, Petrescu A (1995) rhGM-CSF in combination with standard induction chemotherapy for the treatment of acute myeloid leukemia – an overview. In: Nissen NI, Dexter TM (eds) Granulocyte-macrophage colony-stimulating factor (GM-CSF). Current use and future applications. Procceedings from an international symposium, Lucerne, Switzerland, September 9–11 1993. Pennine Press, Cheshire, pp 24–26

Hattersley G, Owens J, Flanagan AM, Chambers TJ (1991) Macrophage colony stimulating factor (M-CSF) is essential for osteoclast formation in vitro. Biochem Biophys Res Commun 177:526–531

Havill AM, Andresen JW, Korach E, Nelson S (1991) Enhancement of endotoxin-induced lethality in the mouse by GM-CSF. Am Rev Respir Dis 143:A236

Hayashida K, Kitamura T, Gorman DM, Arai K, Yokota T, Miyajima A (1990) Molecular cloning of a second subunit of the receptor for human granulocyte-macrophage colony-stimulating factor (GM-CSF): reconstitution of a high-affinity GM-CSF receptor. Proc Natl Acad Sci USA 87:9655–9659

Hayes MP, Wang J, Norcross MA (1995) Regulation of interleukin-12 expression in human monocytes: selective priming by interferon-gamma of lipopolysaccharide-inducible p35 and p40 genes. Blood 86:646–650

Heard JM, Roussel MF, Rettenmier CW, Sherr CJ (1987) Synthesis, post-translational processing, and autocrine transforming activity of a carboxylterminal truncated form of colony stimulating factor-1. Oncogene Res 1:423–440

Heidenreich S, Gong JH, Schmidt A, Nain M, Gemsa D (1989) Macrophage activation by granulocyte/macrophage colony-stimulating factor. Priming for enhanced release of tumor necrosis factor-alpha and prostaglandin E2. J Immunol 143:1198–1205

Held TK, Mielke ME, Unger M, Trautmann M, Cross AS (1996) Kinetics and dose dependence of macrophage colony-stimulating factor-induced proliferation and activation of murine mononuclear phagocytes in situ: differences between lungs, liver, and spleen. J Interferon Cytokine Res 16:159–168

Hengge UR, Brockmeyer NH, Goos M (1992) Granulocyte colony-stimulating factor treatment in AIDS patients. Clin Investig 70:922–926

Henschler R, Lindemann A, Brach MA, Mackensen A, Mertelsmann RH, Herrmann F (1991) Expression of functional receptors for interleukin-6 by human polymorphonuclear leukocytes. Downregulation by granulocyte-macrophage colony-stimulating factor. FEBS Lett 283:47–51

Hensler T, Konig B, Prevost G, Piemont Y, Koller M, Konig W (1994) Leukotriene B4 generation and DNA fragmentation induced by leukocidin from Staphylococcus aureus: protective role of granulocyte-macrophage colony-stimulating factor (GM-CSF) and G-CSF for human neutrophils. Infect Immun 62:2529–2535

Herrmann F, Cannistra SA, Griffin JD (1986) T cell-monocyte interactions in the production of humoral factors regulating human granulopoiesis in vitro. J Immunol 136:2856–2861

Herrmann F, Lindemann A, Mertelsmann R (1990) G-CSF and M-CSF: from molecular biology to clinical application. Biotherapy 2:315–324

Herrmann F, Oster W, Meuer SC, Lindemann A, Mertelsmann RH (1988) Interleukin 1 stimulates T lymphocytes to produce granulocyte-monocyte colony-stimulating factor. J Clin Invest 81:1415–1418

Hirai K, Morita Y, Misaki Y, Ohta K, Takaishi T, Suzuki S, Motoyoshi K, Miyamoto T (1988) Modulation of human basophil histamine release by hemopoietic growth factors. J Immunol 141:3958–3964

Hirokawa M, Kitabayashi A, Niitsu H, Takatsu H, Miura AB (1995) Modulation of allogeneic immune responses by filgrastim (recombinant human granulocyte colony-stimulating factor) in bone marrow transplantation. Int J Hematol 62:235–241

Ho JL, Reed SG, Wick EA, Giordano M (1990) Granulocyte-macrophage and macrophage colony-stimulating factors activate intramacrophage killing of Leishmania mexicana amazonensis. J Infect Dis 162:224–230

Hogasen AK, Hestdal K, Abrahamsen TG (1993) Granulocyte-macrophage colony-stimulating factor, but not macrophage colony-stimulating factor, suppresses basal and lipopolysaccharide-stimulated complement factor production in human monocytes. J Immunol 151:3215–3224

Hollingshead LM, Goa KL (1991) Recombinant granulocyte colony-stimulating factor (rG-CSF). A review of its pharmacological properties and prospective role in neutropenic conditions. Drugs 42:300–330

Hommes DW, Meenan J, Dijkhuizen S, Ten Kate FJ, Tytgat GN, Van Deventer SJ (1996) Efficacy of recombinant granulocyte colony-stimulating factor (rhG-CSF) in experimental colitis. Clin Exp Immunol 106:529–533

Horiguchi J, Warren MK, Kufe D (1987) Expression of the macrophage-specific colony-stimulating factor in human monocytes treated with granulocyte-macrophage colony-stimulating factor. Blood 69:1259–1261

Horiguchi J, Warren MK, Ralph P, Kufe D (1986) Expression of the macrophage specific colony-stimulating factor (CSF-1) during human monocytic differentiation. Biochem Biophys Res Commun 141:924–930

Huebner K, Isobe M, Croce CM, Golde DW, Kaufman SE, Gasson JC (1985) The human gene encoding GM-CSF is at 5q21-q32, the chromosome region deleted in the 5q- anomaly. Science 230:1282–1285

Iacobini M, Palumbo G, Bartolozzi S, Mondaini C, Castello M, Clerico A, Perla FM, Werner B, Digilio G (1995) Superoxide production by neutrophils in children with malignant tumors treated with recombinant human granulocyte colony-stimulating factor. Haematologica 80:13–17

Ibelgaufts H (1992) Lexikon der Zytokine. Medikon Verlag, München

Ichinose Y, Hara N, Ohta M, Aso H, Chikama H, Kawasaki M, Kubota I, Shimizu T, Yagawa K (1990) Recombinant granulocyte colony-stimulating factor and lipopolysaccharide maintain the phenotype of and superoxide anion generation by neutrophils. Infect Immun 58:1647–1652

Iguchi K, Inoue S, Kumar A (1991) Effect of recombinant human granulocyte colony-stimulating factor administration in normal and experimentally infected newborn rats. Exp Hematol 19:352–358

Ihle JN, Keller J, Oroszlan S, Henderson LE, Copeland TD, Fitch F, Prystowsky MB, Goldwasser E, Schrader JW, Palaszynski E, Dy M, Lebel B (1983) Biologic properties of homogenous interleukin 3. J Immunol 131:282–287

Imakawa K, Helmer SD, Nephew KP, Meka CS, Christenson RK (1993) A novel role for GM-CSF: enhancement of pregnancy specific interferon production, ovine trophoblast protein-1. Endocrinology 132:1869–1871

Inaba K, Steinman RM, Pack MW, Aya H, Inaba M, Sudo T, Wolpe S, Schuler G (1992) Identification of proliferating dendritic cell precursors in mouse blood. J Exp Med 175:1157–1167

Ishizaka Y, Motoyoshi K, Hatake K, Saito M, Takaku F, Miura Y (1986) Mode of action of human urinary colony-stimulating factor. Exp Hematol 14:1–8

Itoh N, Yonehara S, Schreurs J, Gorman DM, Maruyama K, Ishii A, Yahara I, Arai K, Miyajima A (1990) Cloning of an interleukin-3 receptor gene: a member of a distinct receptor gene family. Science 247:324–327

Jenkins JK, Arend WP (1993) Interleukin 1 receptor antagonist production in human monocytes is induced by IL-1 alpha, IL-3, IL-4 and GM-CSF. Cytokine 5:407-415

Johnson GR, Gonda TJ, Metcalf D, Hariharan IK, Cory S (1989) A lethal myeloproliferative syndrome in mice transplanted with bone marrow cells infected with a retrovirus expressing granulocyte-macrophage colony stimulating factor. EMBO J 8:441–448

Jones T, Stern A, Lin R (1994) Potential role of granulocyte-macrophage colony-stimulating factor as vaccine adjuvant. Eur J Clin Microbiol Infect Dis 13 Suppl 2:S47–53

Jones TC (1993) The Effects of rhGM-CSF on Macrophage Function. Eur J Cancer 29A:S10–S13

Jung HC, Eckmann L, Yang SK, Panja A, Fierer J, Morzycka-Wroblewska E, Kagnoff MF (1995) A distinct array of proinflammatory cytokines is expressed in human colon epithelial cells in response to bacterial invasion. J Clin Invest 95:55–65

Jyung RW, Wu L, Pierce GF, Mustoe TA (1994) Granulocyte-macrophage colony-stimulating factor and granulocyte colony-stimulating factor: differential action on incisional wound healing [see comments]. Surgery 115:325–334

Kadota J, Hirota M, Tomono K, Senju R, Fukushima K, Dotsu Y, Komori K, Kohno S, Yamaguchi K, Hara K (1990) [Studies on defense effects of recombinant human granulocyte colony-stimulating factor (G-CSF) to infections. II. Priming effect for superoxide production by human neutrophil]. Kansenshogaku Zasshi 64:430-435

Kalra R, Dale D, Freedman M, Bonilla MA, Weinblatt M, Ganser A, Bowman P, Abish S, Priest J, Oseas RS, et al (1995) Monosomy 7 and activating RAS mutations accompany malignant transformation in patients with congenital neutropenia. Blood 86:4579–4586

Kanazawa M, Ishizaka A, Hasegawa N, Suzuki Y, Yokoyama T (1988) Granulocyte colony-stimulating factor does not enhance endotoxin-induced acute lung injury in Guinea pigs. Am Rev Respir Dis 148:1030–1035

Kanz L (1994) PBPC – Aktueller Stand und Perspektiven. In: Hiddemann W (ed) Klinische Anwendung von rh GM-CSF: Neue Möglichkeiten in der Behandlung von Tumorpatienten. Satelliten-Symposium, 11.03.1994. Hamburg,

Kaplan SS, Basford RE, Wing EJ, Shadduck RK (1989) The effect of recombinant human granulocyte macrophage colony-stimulating factor on neutrophil activation in patients with refractory carcinoma. Blood 73:636–638

Kato T, Inaba K, Ogawa Y, Inaba M, Kakihara K, Shimizu S, Ikehara S, Sudo T, Muramatsu S (1990) Granulocyte-macrophage colony-stimulating factor enhances macrophage accessory function in con A-stimulated T-cell proliferation. Cell Immunol 130:490–500

Katoh M, Shirai T, Shikoshi K, Ishii M, Saito M, Kitagawa S (1992) Neutrophil kinetics shortly after initial administration of recombinant human granulocyte colony-stimulating factor: neutrophil alkaline phosphatase activity as an endogenous marker. Eur J Haematol 49:19–24

Katsura Y, Tsuru S, Noritake M, Kayashima S, Wakiyama H, Shinomiya N, Shinomiya M, Rokutanda M (1993) Granulocyte colony-stimulating factor enhances the extracellular emission of reactive oxygen from neutrophils stimulated with formylmethionylleucylphenylalanine. Cell Immunol 148:10–17

Kaufman SE, Di Persio JF, Gasson JC (1989) Effects of human GM-CSF on neutrophil degranulation in vitro. Exp Hematol 17:800–804

Kawakami M, Tsutsumi H, Kumakawa T, Abe H, Hirai M, Kurosawa S, Mori M, Fukushima M (1990) Levels of serum granulocyte colony-stimulating factor in patients with infections. Blood 76:1962–1964

Kawakami M, Tsutsumi H, Kumakawa T, Hirai M, Kurosawa S, Mori M, Fukushima M (1992) Serum granulocyte colony-stimulating factor in patients with repeated infections. Am J Hematol 41:190–193

Kay AB, Ying S, Varney V, Gaga M, Durham SR, Moqbel R, Wardlaw AJ, Hamid Q (1991) Messenger RNA expression of the cytokine gene cluster, interleukin 3 (IL-3), IL-4, IL-5, and granulocyte/macrophage colony-stimulating factor, in allergen-induced late-phase cutaneous reactions in atopic subjects. J Exp Med 173:775–778

Kayashima S, Tsuru S, Hata N, Rokutanda M (1993) Therapeutic effect of granulocyte colony-stimulating factor (G-CSF) on the protection against Listeria infection in SCID mice. Immunology 80:471–476

Keller P, Smalling R (1993) Granulocyte colony stimulating factor: animal studies for risk assessment. Int Rev Exp Pathol 34 :173–188

Kelley KW, Arkins S, Minshall C, Liu Q, Dantzer R (1996) Growth hormone, growth factors and hematopoiesis. Horm Res 45:38–45

Kerrigan DP, Castillo A, Foucar K, Townsend K, Neidhart J (1989) Peripheral blood morphologic changes after high-dose antineoplastic chemotherapy and recombinant human granulocyte colony-stimulating factor administration. Am J Clin Pathol 92:280–285

Kerst JM, de Haas M, van der Schoot CE, Slaper-Cortenbach IC, Kleijer M, von dem Borne AE, van Oers RH (1993b) Recombinant granulocyte colony-stimulating factor administration to healthy volunteers: induction of immunophenotypically and functionally altered neutrophils via an effect on myeloid progenitor cells. Blood 82:3265–3272

Kerst JM, van de Winkel JG, Evans AH, de Haas M, Slaper-Cortenbach IC, de Wit TP, von dem Borne AE, van der Schoot CE, van Oers RH (1993a) Granulocyte colony-stimulating factor induces hFc gamma RI (CD64 antigen)-positive neutrophils via an effect on myeloid precursor cells. Blood 81:1457–1464

Kharazmi A, Nielsen H, Bendtzen K (1988) Modulation of human neutrophil and monocyte chemotaxis and superoxide responses by recombinant TNF-alpha and GM-CSF. Immunobiology 177:363–370

Khwaja A, Carver JE, Linch DC (1992) Interactions of granulocyte-macrophage colony-stimulating factor (CSF), granulocyte CSF, and tumor necrosis factor alpha in the priming of the neutrophil respiratory burst. Blood 79:745–753

Kindler V, Shields J, Ayer D, Benotto W, Mazzei GJ (1990) Granulocyte-macrophage colony stimulating factor and interleukin-3 regulate the production of an interleukin-1 inhibitor by the differentiated AML-193 leukemic cell line. Eur Cytokine Netw 1:169–176

Kindler V, Thorens B, de Kossodo S, Allet B, Eliason JF, Thatcher D, Farber N, Vassalli P (1986) Stimulation of hematopoiesis in vivo by recombinant bacterial murine interleukin 3. Proc Natl Acad Sci USA 83:1001–1005

Kishimoto S, Shimazu W, Izumi T, Shimizu T, Fukuda T, Makino S, Sugiura T, Waku K (1996) Enhanced expression of platelet-activating factor receptor on human eosinophils by interleukin-3, interleukin-5 and granulocyte-macrophage colony-stimulating factor. Int Arch Allergy Immunol 111:63–65

Kita H, Ohnishi T, Okubo Y, Weiler D, Abrams JS, Gleich GJ (1991) Granulocyte/macrophage colony-stimulating factor and interleukin 3 release from human peripheral blood eosinophils and neutrophils. J Exp Med 174:745–748

Kitabayashi A, Hirokawa M, Hatano Y, Lee M, Kuroki J, Niitsu H, Miura AB (1995) Granulocyte colony-stimulating factor downregulates allogeneic immune responses by posttranscriptional inhibition of tumor necrosis factor-alpha production. Blood 86:2220–2227

Kitagawa S, Yuo A, Souza LM, Saito M, Miura Y, Takaku F (1987) Recombinant human granulocyte colony-stimulating factor enhances superoxide release in human granulocytes stimulated by the chemotactic peptide. Biochem Biophys Res Commun 144:1143–1146

Kitamura T, Sato N, Arai K, Miyajima A (1991) Expression cloning of the human IL-3 receptor cDNA reveals a shared beta subunit for the human IL-3 and GM-CSF receptors. Cell 66:1165–1174

Kleinerman ES, Knowles RD, Lachman LB, Gutterman JU (1988) Effect of recombinant granulocyte/macrophage colony-stimulating factor on human monocyte activity in vitro and following intravenous administration. Cancer Res 48:2604–2609

Koeffler HP, Gasson J, Ranyard J, Souza L, Shepard M, Munker R (1987) Recombinant human TNF alpha stimulates production of granulocyte colony-stimulating factor. Blood 70:55–59

Koeffler HP, Gasson J, Tobler A (1988) Transcriptional and posttranscriptional modulation of myeloid colony-stimulating factor expression by tumor necrosis factor and other agents. Mol Cell Biol 8:3432–3438

Kohn FR, Phillips GL, Klingemann HG (1992) Regulation of tumor necrosis factor-alpha production and gene expression in monocytes. Bone Marrow Transplant 9:369–376

Konig B, Konig W (1994) Effect of growth factors on Escherichia coli alpha-hemolysin-induced mediator release from human inflammatory cells: involvement of the signal transduction pathway. Infect Immun 62:2085–2093

Krause SW, Kreutz M, Zenke G, Andreesen R (1992) Developmental regulation of granulocyte-macrophage colony-stimulating factor production during human monocyte-to-macrophage maturation. Ann Hematol 64:190–195

Krishnaswamy G, Smith JK, Srikanth S, Chi DS, Kalbfleisch JH, Huang SK (1996) Lymphoblastoid interferon-alpha inhibits T cell proliferation and expression of eosinophil-activating cytokines. J Interferon Cytokine Res 16:819–827

Kropec A, Lemmen SW, Grundmann HJ, Engels I, Daschner FD (1995) Synergy of simultaneous administration of ofloxacin and granulocyte colony-stimulating factor in killing of Escherichia coli by human neutrophils. Infection 23:298–300

Kruger M, Coorevits L, De Wit TP, Casteels-Van Daele M, Van De Winkel JG, Ceuppens JL (1996b) Granulocyte-macrophage colony-stimulating factor antagonizes the transforming growth factor-beta-induced expression of Fc gamma RIII (CD16) on human monocytes. Immunology 87:162–167

Kruger M, Van de Winkel JG, De Wit TP, Coorevits L, Ceuppens JL (1996a) Granulocyte-macrophage colony-stimulating factor down-regulates CD14 expression on monocytes. Immunology 89:89–95

Kucukcelebi A, Carp SS, Hayward PG, Hui P-S, Cowan WT, Ko F, Cooper DM, Robson MC (1992) Granulocyte-macrophage colony stimulating factor reverses the inhibition of wound contraction caused by bacterial contamination. Wounds 4:241–247

Kushner BH, Cheung NK (1989) GM-CSF enhances 3F8 monoclonal antibody-dependent cellular cytotoxicity against human melanoma and neuroblastoma. Blood 73:1936–1941

Lai CF, Baumann H (1996) Interleukin-1 beta induces production of granulocyte colony-stimulating factor in human hepatoma cells. Blood 87:4143–4148

Lamas AM, Leon OG, Schleimer RP (1991) Glucocorticoids inhibit eosinophil responses to granulocyte-macrophage colony-stimulating factor. J Immunol 147:254–259

Lanfrancone L, Boraschi D, Ghiara P, Falini B, Grignani F, Peri G, Mantovani A, Pelicci PG (1992) Human peritoneal mesothelial cells produce many cytokines (granulocyte colony-stimulating factor [CSF], granulocyte-monocyte-CSF, macrophage-CSF, interleukin-1 [IL-1], and IL-6) and are activated and stimulated to grow by IL-1. Blood 80:2835–2842

Lang CH, Bagby GJ, Dobrescu C, Nelson S, Spitzer JJ (1992a) Effect of granulocyte colony-stimulating factor on sepsis-induced changes in neutrophil accumulation and organ glucose uptake. J Infect Dis 166:336–343

Lang CH, Bagby GJ, Dobrescu C, Nelson S, Spitzer JJ (1992b) Modulation of glucose metabolic response to endotoxin by granulocyte colony-stimulating factor. Am J Physiol 263:R1122–1129

Lang RA, Metcalf D, Cuthbertson RA, Lyons I, Stanley E, Kelso A, Kannourakis G, Williamson DJ, Klintworth GK, Gonda TJ, et al (1987) Transgenic mice expressing a hemopoietic growth factor gene (GM-CSF) develop accumulations of macrophages, blindness, and a fatal syndrome of tissue damage. Cell 51:675–686

Layton JE, Hockman H, Sheridan WP, Morstyn G (1989) Evidence for a novel in vivo control mechanism of granulopoiesis: mature cell-related control of a regulatory growth factor. Blood 74:1303–1307

Lazard T, Perronne C, Cohen Y, Grosset J, Vilde JL, Pocidalo JJ (1993) Efficacy of granulocyte colony-stimulating factor and RU-40555 in combination with clarithromycin against Mycobacterium avium complex infection in C57BL/6 mice. Antimicrob Agents Chemother 37:692–695

Lee MT, Kaushansky K, Ralph P, Ladner MB (1990) Differential expression of M-CSF, G-CSF, and GM-CSF by human monocytes. J Leukoc Biol 47:275–282

Lejeune M, Sariban ECantinieaux,B, Ferster A, Devalkck C, Fondu P (1996) Defective polymorphonuclear leukocyte functions in children receiving chemotherapy for cancer are partially restored by recombinant human granulocyte colony-stimulating factor in vitro. J Infect Dis 174:800–805

Lenhoff S, Olofsson T (1996) Effects of immunosuppressive drugs and antibiotics on GM-CSF and G-CSF secretion in vitro by monocytes, T lymphocytes and endothelial cells. Br J Haematol 95:33–38

Levitt LJ, Nagler A, Lee F, Abrams J, Shatsky M, Thompson D (1991) Production of granulocyte/macrophage-colony-stimulating factor by human natural killer cells. Modulation by the p75 subunit of the interleukin 2 receptor and by the CD2 receptor. J Clin Invest 88:67–75

Liehl E, Hildebrandt J, Lam C, Mayer P (1994) Prediction of the role of granulocyte-macrophage colony-stimulating factor in animals and man from in vitro results. Eur J Clin Microbiol Infect Dis 13 Suppl 2:S9–17

Lieschke GJ, Burgess AW (1992a) Granulocyte colony-stimulating factor and granulocyte-macrophage colony-stimulating factor (2). N Engl J Med 327:99–106

Lieschke GJ, Burgess AW (1992b) Granulocyte colony-stimulating factor and granulocyte-macrophage colony-stimulating factor (1). N Engl J Med 327:28–35

Lieschke GJ, Grail D, Hodgson G, Metcalf D, Stanley E, Cheers C, Fowler KJ, Basu S, Zhan YF, Dunn AR (1994a) Mice lacking granulocyte colony-stimulating factor have chronic neutropenia, granulocyte and macrophage progenitor cell deficiency, and impaired neutrophil mobilization. Blood 84:1737–1746

Lieschke GJ, Stanley E, Grail D, Hodgson G, Sinickas V, Gall JA, Sinclair RA, Dunn AR (1994b) Mice lacking both macrophage- and granulocyte-macrophage colony-stimulating factor have macrophages and coexistent osteopetrosis and severe lung disease. Blood 84:27–35

Liesveld JL, Abboud CN, Looney RJ, Ryan DH, Brennan JK (1988) Expression of IgG Fc receptors in myeloid leukemic cell lines. Effect of colony-stimulating factors and cytokines. J Immunol 140:1527–1533

Liles WC, Huang JE, Llewellyn C, Sen Gupta D, Price TH, Dale DC (1997b) A comparative trial of granulocyte-colony-stimulating factor and dexamethasone, separately and in combination, for the mobilization of neutrophils in the peripheral blood of normal volunteers. Transfusion 37:182–187

Liles WC, Huang JE, van Burik JA, Bowden RA, Dale DC (1997a) Granulocyte colony-stimulating factor administered in vivo augments neutrophil-mediated activity against opportunistic fungal pathogens. J Infect Dis 175:1012–1015

Liles WC, Kiener PA, Ledbetter JA, Aruffo A, Klebanoff SJ (1996) Differential expression of Fas (CD95) and Fas ligand on normal human phagocytes: implications for the regulation of apoptosis in neutrophils. J Exp Med 184:429–440

Liles WC, Rodger ER, Dale DC (1994a) Differential regualtion of human neutrophil surface expression of CD14, CD16, CD 18, and L-selectin following the administration of G-CSF in vivo and in vitro. Clin Res 42:304A

Liles WC, Waltersdorph AM, Klebanoff SJ (1994b) Regulation of apoptosis in human neutrophils: Effects of proinflammatory mediators. Protein kinase inhibitors and antibodies directed against B2-integrins. Clin Res 42:148a

Lindemann A, Herrmann F, Oster W, Haffner G, Meyenburg W, Souza LM, Mertelsmann R (1989a) Hematologic effects of recombinant human granulocyte colony-stimulating factor in patients with malignancy. Blood 74:2644–2651

Lindemann A, Riedel D, Oster W, Meuer SC, Blohm D, Mertelsmann RH, Herrmann F (1988) Granulocyte/macrophage colony-stimulating factor induces interleukin 1 production by human polymorphonuclear neutrophils. J Immunol 140:837–839

Lindemann A, Riedel D, Oster W, Ziegler-Heitbrock HW, Mertelsmann R, Herrmann F (1989b) Granulocyte-macrophage colony-stimulating factor induces cytokine secretion by human polymorphonuclear leukocytes. J Clin Invest 83:1308–1312

Lister PD, Gentry MJ, Preheim LC (1993) Granulocyte colony-stimulating factor protects control rats but not ethanol-fed rats from fatal pneumococcal pneumonia. J Infect Dis 168:922–926

Liu F, Wu HY, Wesselschmidt R, Kornaga T, Link DC (1996) Impaired production and increased apoptosis of neutrophils in granulocyte colony-stimulating factor receptor-deficient mice. Immunity 5:491–501

Liu KY, Akashi K, Harada M, Takamatsu Y, Niho Y (1993) Kinetics of circulating haematopoietic progenitors during chemotherapy-induced mobilization with or without granulocyte colony-stimulating factor. Br J Haematol 84:31–38

Lopez AF, Eglinton JM, Gillis D, Park LS, Clark S, Vadas MA (1989) Reciprocal inhibition of binding between interleukin 3 and granulocyte-macrophage colony-stimulating factor to human eosinophils. Proc Natl Acad Sci USA 86:7022–7026

Lopez AF, Eglinton JM, Lyons AB, Tapley PM, To LB, Park LS, Clark SC, Vadas MA (1990) Human interleukin-3 inhibits the binding of granulocyte-macrophage colony-stimulating factor and interleukin-5 to basophils and strongly enhances their functional activity. J Cell Physiol 145:69–77

Lopez AF, Elliott MJ, Woodcock J, Vadas MA (1992) GM-CSF, IL-3 and IL-5: cross-competition on human haemopoietic cells. Immunol Today 13:495–500

Lopez AF, Nicola NA, Burgess AW, Metcalf D, Battye FL, Sewell WA, Vadas M (1983) Activation of granulocyte cytotoxic function by purified mouse colony-stimulating factors. J Immunol 131:2983–2988

Lopez AF, Williamson DJ, Gamble JR, Begley CG, Harlan JM, Klebanoff SJ, Waltersdorph A, Wong G, Clark SC, Vadas MA (1986) Recombinant human granulocyte-macrophage colony-stimulating factor stimulates in vitro mature human neutrophil and eosinophil function, surface receptor expression, and survival. J Clin Invest 78:1220–1228

Lopez M, Martinache C, Canepa S, Chokri M, Scotto F, Bartholeyns J (1993) Autologous lymphocytes prevent the death of monocytes in culture and promote, as do GM-CSF, IL-3 and M-CSF, their differentiation into macrophages. J Immunol Methods 159:29–38

Lord BI, Molineux G, Pojda Z, Souza LM, Mermod JJ, Dexter TM (1991) Myeloid cell kinetics in mice treated with recombinant interleukin-3, granulocyte colony-stimulating factor (CSF), or granulocyte-macrophage CSF in vivo. Blood 77:2154–2159

Lorenz W, Reimund KP, Weitzel F, Celik I, Kurnatowski M, Schneider C, Mannheim W, Heiske A, Neumann K, Sitter H, et al (1994) Granulocyte colony-stimulating factor prophylaxis before operation protects against lethal consequences of postoperative peritonitis. Surgery 116:925–934

Lu L, Srour EF, Warren DJ, Walker D, Graham CD, Walker EB, Jansen J, Broxmeyer HE (1988b) Enhancement of release of granulocyte- and granulocyte-macrophage colony-stimulating factors from phytohemagglutinin-stimulated sorted subsets of human T lymphocytes by recombinant human tumor necrosis factor-alpha. Synergism with recombinant human IFN-gamma. J Immunol 141:201–207

Lu L, Walker D, Graham CD, Waheed A, Shadduck RK, Broxmeyer HE (1988a) Enhancement of release from MHC class II antigen-positive monocytes of hematopoietic colony stimulating factors CSF-1 and G-CSF by recombinant human tumor necrosis factor-alpha: synergism with recombinant human interferon-gamma. Blood 72:34–41

Lundblad R, Nesland JM, Giercksky KE (1996) Granulocyte colony-stimulating factor improves survival rate and reduces concentrations of bacteria, endotoxin, tumor necrosis factor, and endothelin-1 in fulminant intra-abdominal sepsis in rats. Crit Care Med 24:820–826

Malyak M, Smith Jr. MF, Abel AA, Arend WP (1994) Peripheral blood neutrophil production of interleukin-1 receptor antagonist and interleukin-1 beta. J Clin Immunol 14:20–30

Mandujano JF, D'Souza NB, Nelson S, Summer WR, Beckerman RC, Shellito JE (1995) Granulocyte-macrophage colony stimulating factor and Pneumocystis carinii pneumonia in mice. Am J Respir Crit Care Med 151:1233–1238

Mareni C, Sessarego M, Montera M, Gugazza G, Origone P, D'Amato E, Lerza R, Pistoia V, Scarra GB (1994) Expression and genomic configuration of GM-CSF, IL-3, M-CSF receptor (C-FMS), early growth response gene-1 (EGR-1) and M-CSF genes in primary myelodysplastic syndromes. Leuk Lymphoma 15:135–141

Marth C, Weiss G, Koza A, Reibnegger G, Daxenbichler G, Zeimet AG, Fuchs D, Wachter H, Dapunt O (1994) Increased production of immune activation marker neopterin by colony-stimulating factors in gynecological cancer patients. Int J Cancer 58:20–23

Matsuda S, Shirafuji N, Asano S (1989) Human granulocyte colony-stimulating factor specifically binds to murine myeloblastic NFS-60 cells and activates their guanosine triphosphate binding proteins/adenylate cyclase system. Blood 74:2343–2348

Matsumoto M, Matsubara S, Matsuno T, Tamura M, Hattori K, Nomura H, Ono M, Yokota T (1987) Protective effect of human granulocyte colony-stimulating factor on microbial infection in neutropenic mice. Infect Immun 55:2715–2720

Matsushime H, Roussel MF, Matsushima K, Hishinuma A, Sherr CJ (1991) Cloning and expression of murine interleukin-1 receptor antagonist in macrophages stimulated by colony-stimulating factor 1. Blood 78:616–623

Maurer AB, Ganser A, Buhl R, Seipelt G, Ottmann OG, Mentzel U, Geissler RG, Hoelzer D (1993) Restoration of impaired cytokine secretion from monocytes of patients with myelodysplastic syndromes after in vivo treatment with GM-CSF or IL-3. Leukemia 7:1728–1733

Maurer D, Fischer GF, Felzmann T, Majdic O, Gschwantler E, Hinterberger W, Wagner A, Knapp W (1991) Ratio of complement receptor over Fc-receptor III expression: a sensitive parameter to monitor granulocyte-macrophage colony-stimulating factor effects on neutrophils. Ann Hematol 62:135–140

Mayer P, Lam C, Obenaus H, Liehl E, Besemer J (1987) Recombinant human GM-CSF induces leukocytosis and activates peripheral blood polymorphonuclear neutrophils in nonhuman primates. Blood 70:206–213

Mayer P, Valent P, Schmidt G, Liehl E, Bettelheim P (1989) The in vivo effects of recombinant human interleukin-3: demonstration of basophil differentiation factor, histamine-producing activity, and priming of GM-CSF-responsive progenitors in nonhuman primates. Blood 74:613–621

Mazzei GJ, Bernasconi LM, Lewis C, Mermod JJ, Kindler V, Shaw AR (1990) Human granulocyte-macrophage colony-stimulating factor plus phorbol myristate acetate stimulate a promyelocytic cell line to produce an IL-1 inhibitor. J Immunol 145:585–591

McColl SR, Di Persio JF, Caon AC, Ho P, Naccache PH (1991) Involvement of tyrosine kinases in the activation of human peripheral blood neutrophils by granulocyte-macrophage colony-stimulating factor. Blood 78:1842–1852

McColl SR, Paquin R, Menard C, Beaulieu AD (1992) Human neutrophils produce high levels of the interleukin 1 receptor antagonist in response to granulocyte/macrophage colony-stimulating factor and tumor necrosis factor alpha. J Exp Med 176:593–598

McHugh S, Deighton J, Rifkin I, Ewan P (1996) Kinetics and functional implications of Th1 and Th2 cytokine production following activation of peripheral blood mononuclear cells in primary culture. Eur J Immunol 26:1260–1265

McKenna PJ, Nelson S, Andreson J (1996) Filgrastim (rbuG-CSF) Enhances Ciprofloxacin uptake and Bactericidal Acitivity of Human Neutrophils in Vitro. Am J Respir Crit Care Med 153

McKenzie SE, Kline J, Douglas SD, Polin RA (1993) Enhancement in vitro of the low interferon-gamma production of leukocytes from human newborn infants. J Leukoc Biol 53:691–696

Medlock ES, Kaplan DL, Cecchini M, Ulich TR, del Castillo J, Andresen J (1993) Granulocyte colony-stimulating factor crosses the placenta and stimulates fetal rat granulopoiesis. Blood 81:916–922

Mempel K, Pietsch T, Menzel T, Zeidler C, Welte K (1991) Increased serum levels of granulocyte colony-stimulating factor in patients with severe congenital neutropenia. Blood 77:1919–1922

Metcalf D (1985) The granulocyte-macrophage colony-stimulating factors. Science 229:16–22

Metcalf D (1986) The molecular biology and functions of the granulocyte-macrophage colony-stimulating factors. Blood 67:257–267

Metcalf D (1987) The role of the colony-stimulating factors in resistance to acute infections. Immunol Cell Biol 65:35–43

Metcalf D, Begley CG, Johnson GR, Nicola NA, Lopez AF, Williamson DJ (1986) Effects of purified bacterially synthesized murine multi-CSF (IL-3) on hematopoiesis in normal adult mice. Blood 68:46–57

Metcalf D, Nicola NA (1983) Proliferative effects of purified granulocyte colony-stimulating factor (G-CSF) on normal mouse hemopoietic cells. J Cell Physiol 116:198–206

Metcalf D, Nicola NA (1985) Synthesis by mouse peritoneal cells of G-CSF, the differentiation inducer for myeloid leukemia cells: stimulation by endotoxin, M-CSF and multi-CSF. Leuk Res 9:35–50

Metcalf D, Nicola NA (1992) The clonal proliferation of normal mouse hematopoietic cells: enhancement and suppression by colony-stimulating factor combinations. Blood 79:2861–2866

Metcalf D, Nicola NA, Gough NM, Elliott M, McArthur G, Li M (1992) Synergistic suppression: anomalous inhibition of the proliferation of factor-dependent hemopoietic cells by combination of two colony-stimulating factors. Proc Natl Acad Sci USA 89:2819–2823

Metcalf D, Robb L, Dunn AR, Mifsud S, Di Rago L (1996) Role of granulocyte-macrophage colony-stimulating factor and granulocyte colony-stimulating factor in the development of an acute neutrophil inflammatory response in mice. Blood 88:3755–3764

Miadonna A, Roncarolo MG, Lorini M, Tedeschi A (1993) Inducing and enhancing effects of IL-3, -5, and -6 and GM-CSF on histamine release from human basophils. Clin Immunol Immunopathol 67:210–215

Miles SA, Mitsuyasu RT, Moreno J, Baldwin G, Alton NK, Souza L, Glaspy JA (1991) Combined therapy with recombinant granulocyte colony-stimulating factor and erythropoietin decreases hematologic toxicity from zidovudine. Blood 77:2109–2117

Misago M, Kikuchi M, Tsukada J, Hanamura T, Kamachi S, Eto S (1991) Serum levels of G-CSF, M-CSF and GM-CSF in a patient with cyclic neutropenia [letter]. Eur J Haematol 46:312–313

Mitsuyasu RT (1994) Clinical Uses of Hematopoietic Growth Hormones in HIV-Related Illnesses. Aids Clinical Review 94:189–212

Molloy RG, Holzheimer R, Nestor M, Collins K, Mannick JA, Rodrick ML (1995) Granulocyte-macrophage colony-stimulating factor modulates immune function and improves survival after experimental thermal injury. Br J Surg 82:770–776

Montgomery B, Bianco JA, Jacobsen A, Singer JW (1991) Localization of transfused neutrophils to site of infection during treatment with recombinant human granulocyte-macrophage colony-stimulating factor and pentoxifylline [letter]. Blood 78:533–534

Mooney DP, Gamelli RL, O'Reilly M, Hebert JC (1988) Recombinant human granulocyte colony-stimulating factor and Pseudomonas burn wound sepsis. Arch Surg 123:1353–1357

Moore KJ, Matlashewski G (1994) Intracellular infection by Leishmania donovani inhibits macrophage apoptosis. J Immunol 152:2930–2937

Moore MA (1991) The clinical use of colony stimulating factors. Annu Rev Immunol 9:159–191

Morrissey PJ, Bressler L, Park LS, Alpert A, Gillis S (1987) Granulocyte-macrophage colony-stimulating factor augments the primary antibody response by enhancing the function of antigen-presenting cells. J Immunol 139:1113–1119

Morstyn G, Campbell L, Souza LM, Alton NK, Keech J, Green M, Sheridan W, Metcalf D, Fox R (1988) Effect of granulocyte colony stimulating factor on neutropenia induced by cytotoxic chemotherapy. Lancet 1:667–672

Morstyn G, Lieschke GJ, Sheridan W, Layton J, Cebon J (1989) Pharmacology of the colony-stimulating factors. Trends Pharmacol Sci 10:154–159

Moses AV, Williams S, Heneveld ML, Strussenberg J, Rarick M, Loveless M, Bagby G, Nelson JA (1996) Human immunodeficiency virus infection of bone marrow endothelium reduces induction of stromal hematopoietic growth factors. Blood 87:919–925

Motoyoshi K, Yoshida K, Hatake K, Saito M, Miura Y, Yanai N, Yamada M, Kawashima T, Wong GG, Temple PA, et al (1989) Recombinant and native human urinary colony-stimulating factor directly augments granulocytic and granulocyte-macrophage colony-stimulating factor production of human peripheral blood monocytes. Exp Hematol 17:68–71

Munker R, Gasson J, Ogawa M, Koeffler HP (1986) Recombinant human TNF induces production of granulocyte-monocyte colony-stimulating factor. Nature 323:79–82

Murata K, Watanabe T, Yamashita T, Gon S, Sendo F (1995) Modulation of rat neutrophil functions by administration of granulocyte colony-stimulating factor. J Leukoc Biol 57:250–256

Murray HW, Cervia JS, Hariprashed J, Taylor AP, Stoeckle MY, Hockman H (1995) Effect of granulocyte-macrophage colony-stimulating factor in experimental Visceral Leishmaniasis. J Clin Invest 95:1183–92

Nagata M, Sedgwick JB, Busse WW (1995) Differential effects of granulocyte-macrophage colony-stimulating factor on eosinophil and neutrophil superoxide anion generation. J Immunol 155:4948–4954

Nagata S, Tsuchiya M, Asano S, Kaziro Y, Yamazaki T, Yamamoto O, Hirata Y, Kubota N, Oheda M, Nomura H, et al (1986) Molecular cloning and expression of cDNA for human granulocyte colony-stimulating factor. Nature 319:415–418

Nakada T, Sato H, Inoue F, Mizorogi F, Nagayama K, Tanaka T (1996) The production of colony-stimulating factors by thyroid carcinoma is associated with marked neutrophilia and eosinophilia. Intern Med 35:815–820

Nakamura Y, Azuma M, Okano Y, Sano T, Takahashi T, Ohmoto Y, Sone S (1996) Upregulatory effects of Interleukin-4 and Interleukin-13 but not Interleukin-10 on granulocyte/macrophage colony-stimulating factor production by human bronchial epithelial cells. Am J Respir Cell Mol Biol 15:680–7

Nash RA, Burstein SA, Storb R, Yang W, Abrams K, Appelbaum FR, Boone T, Deeg HJ, Durack LD, Schuening FG (1995) Thrombocytopenia in dogs induced by granulocyte-macrophage colony-stimulating factor: increased destruction of circulating platelets. Blood 86:1765–1775

Nathan CF (1989) Respiratory burst in adherent human neutrophils: triggering by colony-stimulating factors CSF-GM and CSF-G. Blood 73:301–306

Nelson S (1994) Role of granulocyte colony-stimulating factor in the immune response to acute bacterial infection in the nonneutropenic host: an overview. Clin Infect Dis 18 Suppl 2:S197–204

Nelson S, Daifuku R, Andresen J (1994) Use of Filgrastim (r-metHu-G-CSF) in Infectious Diseases. In: Morstyn G, Dexter G (eds) Filgrastim in Clinical Practice. Dekker, New York, pp 253–266

Nelson S, Summer W, Bagby G, Nakamura C, Stewart L, Lipscomb G, Andresen J (1991) Granulocyte colony-stimulating factor enhances pulmonary host defenses in normal and ethanol-treated rats. J Infect Dis 164:901–906

Nemunaitis J, Meyers JD, Buckner CD, Shannon-Dorcy K, Mori M, Shulman H, Bianco JA, Higano CS, Groves E, Storb R, et al (1991) Phase I trial of recombinant human macrophage colony-stimulating factor in patients with invasive fungal infections. Blood 78:907–913

Neuman E, Huleatt JW, Jack RM (1990) Granulocyte-macrophage colony-stimulating factor increases synthesis and expression of CR1 and CR3 by human peripheral blood neutrophils. J Immunol 145:3325–3332

Nicola NA (1989) Hemopoietic cell growth factors and their receptors. Annu Rev Biochem 58:45–77

Nicola NA, Metcalf D (1985) Binding of 125I-labeled granulocyte colony-stimulating factor to normal murine hemopoietic cells. J Cell Physiol 124:313–321

Nicola NA, Vadas MA, Lopez AF (1986) Down-modulation of receptors for granulocyte colony-stimulating factor on human neutrophils by granulocyte-activating agents. J Cell Physiol 128:501–509

Nimer SD, Gates MJ, Koeffler HP, Gasson JC (1989) Multiple mechanisms control the expression of granulocyte-macrophage colony-stimulating factor by human fibroblasts. J Immunol 143:2374–2377

Nissen C, Carbonare VD, Moser Y (1994) In vitro comparison of the biological potency of glycosylated versus nonglycosylated rG-CSF. Drug Invest 7:346–52

Nohava K, Malipiero U, Frei K, Fontana A (1992) Neurons and neuroblastoma as a source of macrophage colony-stimulating factor. Eur J Immunol 22:2539–2545

Novales JS, Salva AM, Modanlou HD, Kaplan DL, del Castillo J, Andersen J, Medlock ES (1993) Maternal administration of granulocyte colony-stimulating factor improves neonatal rat survival after a lethal group B streptococcal infection. Blood 81:923–927

Ogawa M (1993) Differentiation and proliferation of hematopoietic stem cells. Blood 81:2844–2853

Oh-eda M, Hasegawa M, Hattori K, Kuboniwa H, Kojima T, Orita T, Tomonou K, Yamazaki T, Ochi N (1990) O-linked sugar chain of human granulocyte colony-stimulating factor protects it against polymerization and denaturation allowing it to retain its biological activity. J Biol Chem 265:11432–11435

Ohno K, Suzumura A, Sawada M, Marunouchi T (1990) Production of granulocyte/macrophage colony-stimulating factor by cultured astrocytes. Biochem Biophys Res Commun 169:719–724

Ohsaka A, Kitagawa S, Sakamoto S, Miura Y, Takanashi N, Takaku F, Saito M (1989) In vivo activation of human neutrophil functions by administration of recombinant human granulocyte colony-stimulating factor in patients with malignant lymphoma. Blood 74:2743–2748

Ohsaka A, Saionji K, Sato N, Mori T, Ishimoto K, Inamatsu T (1993) Granulocyte colony-stimulating factor down-regulates the surface expression of the human leucocyte adhesion molecule-1 on human neutrophils in vitro and in vivo. Br J Haematol 84:574–580

Ohta Y, Tamura S, Tezuka E, Sugawara M, Imai S, Tanaka H (1988) Autoimmune MRL/lpr mice sera contain IgG with interleukin 3-like activity. J Immunol 140:520–525

Ohtoshi T, Takizawa H, Sakamaki C, Ishii A, Hirai K, Morita Y, Ito K, Ohta K, Mano K, Suzuki S (1994) [Cytokine production by human airway epithelial cells and its modulation]. Nippon Kyobu Shikkan Gakkai Zasshi 32 Suppl:65–72

Okuda K, Sanghera JS, Pelech SL, Kanakura Y, Hallek M, Griffin JD, Druker BJ (1992) Granulocyte-macrophage colony-stimulating factor, interleukin-3, and steel factor induce rapid tyrosine phosphorylation of p42 and p44 MAP kinase. Blood 79:2880–2887

Omori F, Okamura S, Shimoda K, Otsuka T, Harada M, Niho Y (1992) Levels of human serum granulocyte colony-stimulating factor and granulocyte-macrophage colony-stimulating factor under pathological conditions. Biotherapy 4:147–153

Ono M, Matsumoto M, Matsubara S, Tomioka S, Asano S (1988) Protective effect of human granulocyte colony-stimulating factor on bacterial and fungal infections in neutropenic mice. Behring Inst Mitt ☒:216–221

O'Reilly M, Silver GM, Greenhalgh DG, Gamelli RL, Davis JH, Hebert JC (1992) Treatment of intra-abdominal infection with granulocyte colony-stimulating factor. J Trauma 33:679–682

Oster W, Brach MA, Gruss HJ, Mertelsmann R, Herrmann F (1992) Interleukin-1 beta. Blood 79:1260–1265

Oster W, Frisch J, Nicolay U, Schulz G (1991) Interleukin-3. Biologic effects and clinical impact. Cancer 67:2712–2717

Oster W, Lindemann A, Horn S, Mertelsmann R, Herrmann F (1987) Tumor necrosis factor (TNF)-alpha but not TNF-beta induces secretion of colony stimulating factor for macrophages (CSF-1) by human monocytes. Blood 70:1700–1703

Oster W, Lindemann A, Mertelsmann R, Herrmann F (1989b) Granulocyte-macrophage colony-stimulating factor (CSF) and multilineage CSF recruit human monocytes to express granulocyte CSF. Blood 73:64–67

Oster W, Lindemann A, Mertelsmann R, Herrmann F (1989c) Production of macrophage-, granulocyte-, granulocyte-macrophage- and multi-colony-stimulating factor by peripheral blood cells. Eur J Immunol 19:543–547

Oster W, Mertelsmann R, Herrmann F (1989a) Role of colony-stimulating factors in the biology of acute myelogenous leukemia. Int J Cell Cloninthe 7:13–29

Otsuka T, Miyajima A, Brown N, Otsu K, Abrams J, Saeland S, Caux C, de Waal Malefijt R, de Vries J, Meyerson P, et al. (1988) Isolation and characterization of an expressible cDNA encoding human IL-3. Induction of IL-3 mRNA in human T cell clones. J Immunol 140:2288–2295

Owen Jr. WF, Rothenberg ME, Silberstein DS, Gasson JC, Stevens RL, Austen KF, Soberman RJ (1987) Regulation of human eosinophil viability, density, and function by granulocyte/macrophage colony-stimulating factor in the presence of 3T3 fibroblasts. J Exp Med 166:129–141

Pan L, Delmonte Jr. J, Jalonen CK, Ferrara JL (1995) Pretreatment of donor mice with granulocyte colony-stimulating factor polarizes donor T lymphocytes toward type-2 cytokine production and reduces severity of experimental graft-versus-host disease. Blood 86:4422–4429

Panterne B, Hatzfeld A, Sansilvestri P, Cardoso A, Monier M, Batard P, Hatzfeld J (1996) IL-3, GM-CSF and CSF-1 modulate c-fms mRNA more rapidly in human early monocytic progenitors than in mature or transformed monocytic cells. J Cell Science 109:1795–1801

Perkins RC, Vadhan-Raj S, Scheule RK, Hamilton R, Holian A (1993) Effects of continuous high dose rhGM-CSF infusion on human monocyte activity. Am J Hematol 43:279–285

Peters WP, Stuart A, Affronti ML, Kim CS, Coleman RE (1988) Neutrophil migration is defective during recombinant human granulocyte-macrophage colony-stimulating factor infusion after autologous bone marrow transplantation in humans. Blood 72:1310–1315

Piguet PF, Grau GE, de Kossodo S (1993) Role of granulocyte-macrophage colony-stimulating factor in pulmonary fibrosis induced in mice by bleomycin. Exp Lung Res 19:579–587

Pitrak DL, Tsai HC, Mullane KM, Sutton SH, Stevens P (1996) Accelerated neutrophil apoptosis in the acquired immunodeficiency syndrome. J Clin Invest 98:2714–2719

Plaut M, Pierce JH, Watson CJ, Hanley-Hyde J, Nordan RP, Paul WE (1989) Mast cell lines produce lymphokines in response to cross-linkage of Fc epsilon RI or to calcium ionophores. Nature 339:64–67

Pojda Z, Molineux G, Dexter TM (1989) Effects of long-term in vivo treatment of mice with purified murine recombinant GM-CSF. Exp Hematol 17:1100–1104

Polak-Wyss A (1991a) Protective effect of human granulocyte colony-stimulating factor (hG-CSF) on Cryptococcus and Aspergillus infections in normal and immunosuppressed mice. Mycoses 34:205–215

Polak-Wyss A (1991b) Protective effect of human granulocyte colony stimulating factor (hG-CSF) on Candida infections in normal and immunosuppressed mice. Mycoses 34:109–118

Pollard JW, Bartocci A, Arceci R, Orlofsky A, Ladner MB, Stanley ER (1987) Apparent role of the macrophage growth factor, CSF-1, in placental development. Nature 330:484–486

Pollmächer T, Korth C, Schreiber W, Hermann D, Mullington J (1996) Effects of repeated administration of granulocyte colony-stimulating factor (G-CSF) on neutrophil counts, plasma cytokine, and cytokine receptor levels. Cytokine 8:799–803

Povolny BT, Lee MY (1993) The role of recombinant human M-CSF, IL-3, GM-CSF and calcitriol in clonal development of osteoclast precursors in primate bone marrow. Exp Hematol 21:532–537

Price TH, Chatta GS, Dale DC (1996) Effect of recombinant granulocyte colony-stimulating factor on neutrophil kinetics in normal young and elderly humans. Blood 88:335–340

Quesniaux VF, Wehrli S, Ziegler I, Legendre B, Wishart W, Fagg B, Schreier MH, Nissen C (1992) Human serum stimulates the production of G-CSF, IL-1, IL-6 and IL-8 by human peripheral blood leucocytes. Br J Haematol 82:6–12

Ralph P, Nacy CA, Meltzer MS, Williams N, Nakoinz I, Leonard EJ (1983) Colony-stimulating factors and regulation of macrophage tumoricidal and microbicidal activities. Cell Immunol 76:10–21

Ralph P, Nakoinz I (1987) Stimulation of macrophage tumoricidal activity by the growth and differentiation factor CSF-1. Cell Immunol 105:270–279

Rambaldi A, Young DC, Griffin JD (1987) Expression of the M-CSF (CSF-1) gene by human monocytes. Blood 69:1409–1413

Randow F, Döcke W-D, Bundschuh DS, Hartung T, Wendel A, Volk H-D (1997) In vitro prevention and reversal of lipopolysaccharide desensitization by IFN-γ, IL-12, and granulocyte-macrophage colony-stimulating factor. J Immun 158:2911–2918

Rapoport AP, Abboud CN, Di Persio JF (1992) Granulocyte-macrophage colony-stimulating factor (GM-CSF) and granulocyte colony-stimulating factor (G-CSF): receptor biology, signal transduction, and neutrophil activation. Blood Rev 6:43–57

Re F, Mengozzi M, Muzio M, Dinarello CA, Mantovani A, Colotta F (1993) Expression of interleukin-1 receptor antagonist (IL-1ra) by human circulating polymorphonuclear cells. Eur J Immunol 23:570–573

Reed SG, Grabstein KH, Pihl DL, Morrissey PJ (1990) Recombinant granulocyte-macrophage colony-stimulating factor restores deficient immune responses in mice with chronic Trypanosoma cruzi infections. J Immunol 145:1564–1570

Reed SG, Nathan CF, Pihl DL, Rodricks P, Shanebeck K, Conlon PJ, Grabstein KH (1987) Recombinant granulocyte/macrophage colony-stimulating factor activates macrophages to inhibit Trypanosoma cruzi and release hydrogen peroxide. Comparison with interferon gamma. J Exp Med 166:1734–1746

Repp R, Valerius T, Sendler A, Gramatzki M, Iro H, Kalden JR, Platzer E (1991) Neutrophils express the high affinity receptor for IgG (Fc gamma RI, CD64) after in vivo application of recombinant human granulocyte colony-stimulating factor. Blood 78:885–889

Repp R, Valerius T, Wieland G, Becker W, Steininger H, Deo Y, Helm G, Gramatzki M, Van de Winkel JG, Lang N, et al (1995) G-CSF-stimulated PMN in immunotherapy of breast cancer with a bispecific antibody to Fc gamma RI and to HER-2/neu (MDX-210). J Hematother 4:415–421

Richter J, Andersson T, Olsson I (1989) Effect of tumor necrosis factor and granulocyte/macrophage colony-stimulating factor on neutrophil degranulation. J Immunol 142:3199–3205

Roberts PJ, Devereux S, Pilkington GR, Linch DC (1990) Fc gamma RII-mediated superoxide production by phagocytes is augmented by GM-CSF without a change in Fc gamma RII expression. J Leukoc Biol 48:247–257

Robins HI, Kutz M, Wiedemann GJ, Katschinski DM, Paul D, Grosen E, Tiggelaar CL, Spriggs D, Gillis W, d'Oleire F (1995) Cytokine induction by 41.8 degrees C whole body hyperthermia. Cancer Lett 97:195–201

Robison RL, Myers LA (1993;34) Preclinical safety assessment of recombinant human GM-CSF in rhesus monkeys. Int Rev Exp Pathol 34 Pt A:149–172

Robson M, Kucukcelebi A, Carp SS, Hayward PG, Hui PS, Cowan WT, Ko F, Cooper DM (1994) Effects of granulocyte-macrophage colony-stimulating factor on wound contraction. Eur J Clin Microbiol Infect Dis 13 Suppl 2:S41–S46

Roilides E, Holmes A, Blake C, Pizzo PA, Walsh TJ (1995) Effects of granulocyte colony-stimulating factor and interferon-gamma on antifungal activity of human polymorphonuclear neutrophils against pseudohyphae of different medically important Candida species. J Leukoc Biol 57:651–656

Roilides E, Pizzo PA (1992) Modulation of host defenses by cytokines: evolving adjuncts in prevention and treatment of serious infections in immunocompromised hosts. Clin Infect Dis 15:508–524

Roilides E, Pizzo PA (1993) Biologicals and hematopoietic cytokines in prevention or treatment of infections in immunocompromised hosts. Hematol Oncol Clin North Am 7:841–864

Roilides E, Uhlig K, Venzon D, Pizzo PA, Walsh TJ (1992) Neutrophil oxidative burst in response to blastoconidia and pseudohyphae of Candida albicans: augmentation by granulocyte colony-stimulating factor and interferon-gamma. J Infect Dis 166:668–673

Roilides E, Uhlig K, Venzon D, Pizzo PA, Walsh TJ (1993) Enhancement of oxidative response and damage caused by human neutrophils to Aspergillus fumigatus hyphae by granulocyte colony-stimulating factor and gamma interferon. Infect Immun 61:1185–1193

Roilides E, Walsh TJ, Pizzo PA, Rubin M (1991) Granulocyte colony-stimulating factor enhances the phagocytic and bactericidal activity of normal and defective human neutrophils. J Infect Dis 163:579–583

Ross HJ, Koeffler HP (1992) Interactions between TNF and GM-CSF. In: Beutler B (ed) Tumor necrosis factors: the molecules and their emerging role in medicine. Raven Press, New York, pp 179–96

Rottem M, Hull G, Metcalfe DD (1994) Demonstration of differential effects of cytokines on mast cells derived from murine bone marrow and peripheral blood mononuclear cells. Exp Hematol 22:1147–1155

Ruef C, Coleman DL (1990) Granulocyte-macrophage colony-stimulating factor: pleiotropic cytokine with potential clinical usefulness. Rev Infect Dis 12:41–62

Ruef C, Coleman DL (1991) [GM-CSF and G-CSF: cytokines in clinical application]. Schweiz Med Wochenschr 121:397–412

Sadeghi R, Hawrylowicz CM, Chernajovsky Y, Feldmann M (1992) Synergism of glucocorticoids with granulocyte macrophage colony stimulating factor (GM-CSF) but not interferon gamma (IFN-gamma) or interleukin-4 (IL-4) on induction of HLA class II expression on human monocytes. Cytokine 4:287–297

Sagawa K, Mochizuki M, Sugita S, Nagai K, Sudo T, Itoh K (1996) Suppression by IL-10 and IL-4 of cytokine production induced by two-way autologous mixed lymphocyte reaction. Cytokine 8:501–506

Sakurai T, Suzu S, Yamada M, Yanai N, Kawashima T, Hatake K, Takaku F, Motoyoshi K (1994) Induction of tumor necrosis factor in mice by recombinant human macrophage colony-stimulating factor. Jpn J Cancer Res 85:80–85

Sallerfors B (1994) Endogenous production and peripheral blood levels of granulocyte-macrophage (GM-) and granulocyte (G-) colony-stimulating factors. Leuk Lymphoma 13:235–247

Sallerfors B, Olofsson T (1991) Granulocyte-macrophage colony-stimulating factor (GM-CSF) and granulocyte colony-stimulating factor (G-CSF) in serum during induction treatment of acute leukaemia. Br J Haematol 78:343–351

Sallerfors B, Olofsson T (1992) Granulocyte-macrophage colony-stimulating factor (GM-CSF) and granulocyte colony-stimulating factor (G-CSF) secretion by adherent monocytes measured by quantitative immunoassays. Eur J Haematol 49:199–207

San Miguel JF, Hernandez MD, Gonzalez M, Lopez-Berges MC, Caballero MD, Vazquez L, Orfao A, Nieto MJ, Corral M, del Canizo MC (1996) A randomized study comparing the effect of GM-CSF and G-CSF on immune reconstitution after autologous bone marrow transplantation. Br J Haematol 94:140–147

Sanjar S, Smith D, Kings MA, Morley J (1990) Pretreatment with rh-GMCSF, but not rh-IL3, enhances PAF-induced eosinophil accumulation in guinea-pig airways. Br J Pharmacol 100:399–400

Sartorelli KH, Silver GM, Gamelli RL (1991) The effect of granulocyte colony-stimulating factor (G-CSF) upon burn-induced defective neutrophil chemotaxis. J Trauma 31:523–9; discussion 529–530

Sato N, Shimizu H (1993) Granulocyte-colony stimulating factor improves an impaired bactericidal function in neutrophils from STZ-induced diabetic rats. Diabetes 42:470–473

Scarffe JH (1991) Emerging clinical uses for GM-CSF. Eur J Cancer 27:1493–504

Schlimok G, Fackler-Schwalbe I, Lindemann F, Gruber R, Flieger D, Heiss M, Reithmüller G (1995) GM-CSF in gastric cancer: monitoring of micrometastatic tumour cells in bone marrow. In: Nissen NI, Dexter TM (eds) Granulocyte-macrophage colony-stimulating factor (GM-CSF). Current use and future applications. Procceedings from an international symposium, Lucerne, Switzerland, September 9–11 1993. Pennine Press, Cheshire, pp 20–24

Schuitemaker H, Kootstra NA, van Oers MH, van Lambalgen R, Tersmette M, Miedema F (1990) Induction of monocyte proliferation and HIV expression by IL-3 does not interfere with anti-viral activity of zidovudine. Blood 76:1490–1493

Schuurman B, Heuff G, Beelen RH, Meyer S (1994) Enhanced killing capacity of human Kupffer cells after activation with human granulocyte/macrophage-colony-stimulating factor and interferon gamma. Cancer Immunol Immunother 39:179–184

Schwab G, Hecht T (1994) Recombinant methionyl granulocyte colony-stimulating factor (filgrastim): a new dimension in immunotherapy. Ann Hematol 69:1–9

Sedgwick JB, Quan SF, Calhoun WJ, Busse WW (1995) Effect of interleukin-5 and granulocyte-macrophage colony stimulating factor on in vitro eosinophil function: comparison with airway eosinophils. J Allergy Clin Immunol 96:375–385

Seelentag W, Mermod JJ, Vassalli P (1989) Interleukin 1 and tumor necrosis factor-alpha additively increase the levels of granulocyte-macrophage and granulocyte colony-stimulating factor (CSF) mRNA in human fibroblasts. Eur J Immunol 19:209–212

Seelentag WK, Mermod JJ, Montesano R, Vassalli P (1987) Additive effects of interleukin 1 and tumour necrosis factor-alpha on the accumulation of the three granulocyte and macrophage colony-stimulating factor mRNAs in human endothelial cells. EMBO J 6:2261–2265

Shadle PJ, Allen JI, Geier MD, Koths K (1989) Detection of endogenous macrophage colony-stimulating factor (M-CSF) in human blood. Exp Hematol 17:154–159

Shannon MF, Coles LS, Fielke RK, Goodall GJ, Lagnado CA, Vadas MA (1992) Three essential promoter elements mediate tumour necrosis factor and interleukin-1 activation of the granulocyte-colony stimulating factor gene. Growth Factors 7:181–193

Shibata Y, Bjorkman DR, Schmidt M, Oghiso Y, Volkman A (1994) Macrophage colony-stimulating factor-induced bone marrow macrophages do not synthesize or release prostaglandin E2. Blood 83:3316–3323

Shibata Y, McCaffrey PG, Sato H, Oghiso Y (1990) Calcium ionophore but not phorbol ester promotes eicosanoids release by proliferating interleukin-3-dependent bone marrow cells. Blood 76:1586–1592

Shieh JH, Peterson RH, Moore MA (1991) Modulation of granulocyte colony-stimulating factor receptors on murine peritoneal exudate macrophages by tumor necrosis factor-alpha. J Immunol 146:2648–2653

Shields J, Bernasconi LM, Benotto W, Shaw AR, Mazzei GJ (1990) Production of a 26,000-dalton interleukin 1 inhibitor by human monocytes is regulated by granulocyte-macrophage colony-stimulating factor. Cytokine 2:122–128

Shimoda K, Okamura S, Harada N, Kondo S, Okamura T, Niho Y (1993) Identification of a functional receptor for granulocyte colony-stimulating factor on platelets. J Clin Invest 91:1310–1313

Shirafuji N, Matsuda S, Ogura H, Tani K, Kodo H, Ozawa K, Nagata S, Asano S, Takaku F (1990) Granulocyte colony-stimulating factor stimulates human mature neutrophilic granulocytes to produce interferon-alpha. Blood 75:17–19

Shitara T, Yugami S, Ijima H, Sotomatu M, Kuroume T (1994) Cytokine profile during high-dose rhG-CSF therapy in severe congenital neutropenia. Am J Hematol 45:58–62

Sica S, Rutella S, Di Mario A, Salutari P, Rumi C, Ortu la Barbera E, Etuk B, Menichella G, D'Onofrio G, Leone G (1996) rhG-CSF in healthy donors: mobilization of peripheral hemopoietic progenitors and effect on peripheral blood leukocytes. J Hematother 5:391–397

Sieff CA, Emerson SG, Donahue RE, Nathan DG, Wang EA, Wong GG, Clark SC (1985) Human recombinant granulocyte-macrophage colony-stimulating factor: a multilineage hematopoietin. Science 230:1171–1173

Sieff CA, Niemeyer CM, Nathan DG, Ekern SC, Bieber FR, Yang YC, Wong G, Clark SC (1987b) Stimulation of human hematopoietic colony formation by recombinant gibbon multi-colony-stimulating factor or interleukin 3. J Clin Invest 80:818–823

Sieff CA, Tsai S, Faller DV (1987a) Interleukin 1 induces cultured human endothelial cell production of granulocyte-macrophage colony-stimulating factor. J Clin Invest 79:48–51

Silberstein DS, Owen WF, Gasson JC, Di Persio JF, Golde DW, Bina JC, Soberman R, Austen KF, David JR (1986) Enhancement of human eosinophil cytotoxicity and

leukotriene synthesis by biosynthetic (recombinant) granulocyte-macrophage colony-stimulating factor. J Immunol 137:3290–3294
Silver GM, Gamelli RL, O'Reilly M (1989) The beneficial effect of granulocyte colony-stimulating factor (G-CSF) in combination with gentamicin on survival after Pseudomonas burn wound infection. Surgery 106:452–455; discussion 455–456
Simms HH, D'Amico R (1994) Granulocyte colony-stimulating factor reverses septic shock-induced polymorphonuclear leukocyte dysfunction. Surgery 115:85–93
Sisson SD, Dinarello CA (1988) Production of interleukin-1 alpha, interleukin-1 beta and tumor necrosis factor by human mononuclear cells stimulated with granulocyte-macrophage colony-stimulating factor [see comments]. Blood 72:1368–1374
Smith JA (1994) Neutrophils, host defense, and inflammation: a double-edged sword. J Leuko Biol 56:672–686
Smith PD, Lamerson CL, Banks SM, Saini SS, Wahl LM, Calderone RA, Wahl SM (1990a) Granulocyte-macrophage colony-stimulating factor augments human monocyte fungicidal activity for Candida albicans. J Infect Dis 161:999–1005
Smith PD, Lamerson CL, Wong HL, Wahl LM, Wahl SM (1990b) Granulocyte-macrophage colony-stimulating factor stimulates human monocyte accessory cell function. J Immunol 144:3829–3834
Smith RJ, Justen JM, Sam LM (1990c) Recombinant human granulocyte-macrophage colony-stimulating factor induces granule exocytosis from human polymorphonuclear neutrophils. Inflammation 14:83–92
Smith WB, Gamble JR, Vadas MA (1994) The role of granulocyte-macrophage and granulocyte colony-stimulating factors in neutrophil transendothelial migration: comparison with interleukin-8 [see comments]. Exp Hematol 22:329–334
Smith WS, Sumnicht GE, Sharpe RW, Samuelson D, Millard FE (1995) Granulocyte colony-stimulating factor versus placebo in addition to penicillin G in a randomized blinded study of gram-negative pneumonia sepsis: analysis of survival and multisystem organ failure. Blood 86:1301–1309
Snoeck HW, Lardon F, Lenjou M, Nys G, Van Bockstaele DR, Peetermans ME (1993) Interferon-gamma and interleukin-4 reciprocally regulate the production of monocytes/macrophages and neutrophils through a direct effect on committed monopotential bone marrow progenitor cells. Eur J Immunol 23:1072–1077
Socinski MA, Cannistra SA, Sullivan R, Elias A, Antman K, Schnipper L, Griffin JD (1988) Granulocyte-macrophage colony-stimulating factor induces the expression of the CD11b surface adhesion molecule on human granulocytes in vivo. Blood 72:691–697
Souza L (1990) CSF in vivo: effects on hematopoiesis. Int J Cell Cloning 8 Suppl 1:227; discussion 227–228
Souza LM, Boone TC, Gabrilove J, Lai PH, Zsebo KM, Murdock DC, Chazin VR, Bruszewski J, Lu H, Chen KK, et al (1986) Recombinant human granulocyte colony-stimulating factor: effects on normal and leukemic myeloid cells. Science 232:61–65
Spertini O, Kansas GS, Munro JM, Griffin JD, Tedder TF (1991) Regulation of leukocyte migration by activation of the leukocyte adhesion molecule-1 (LAM-1) selectin. Nature 349:691–694
Spivak JL, Smith RR, Ihle JN (1985) Interleukin 3 promotes the in vitro proliferation of murine pluripotent hematopoietic stem cells. J Clin Invest 76:1613–1621
Stanley E, Lieschke GJ, Grail D, Metcalf D, Hodgson G, Gall JA, Maher DW, Cebon J, Sinickas V, Dunn AR (1994) Granulocyte/macrophage colony-stimulating factor-

deficient mice show no major perturbation of hematopoiesis but develop a characteristic pulmonary pathology. Proc Natl Acad Sci USA 91:5592–5596

Steward WP (1993) Granulocyte and granulocyte-macrophage colony-stimulating factors [see comments]. Lancet 342:153–157

Stockmeyer B, Valerius T, Repp R, Heijnen IA, Buhring HJ, Deo YM, Kalden JR, Gramatzki M, van de Winkel JG (1997) Preclinical studies with Fc(gamma)R bispecific antibodies and granulocyte colony-stimulating factor-primed neutrophils as effector cells against HER-2/neu overexpressing breast cancer. Cancer Res 57:696–701

Strassmann G, Bertolini DR, Kerby SB, Fong M (1991) Regulation of murine mononuclear phagocyte inflammatory products by macrophage colony-stimulating factor. Lack of IL-1 and prostaglandin E2 production and generation of a specific IL-1 inhibitor. J Immunol 147:1279–1285

Sullivan GW, Carper HT, Mandell GL (1993) The effect of three human recombinant hematopoietic growth factors (granulocyte-macrophage colony-stimulating factor, granulocyte colony-stimulating factor, and interleukin-3) on phagocyte oxidative activity. Blood 81:1863–1870

Sullivan GW, Gelrud AK, Carper HT, Mandell GL (1996) Interaction of Tumor Necrosis Factor-alpha and Granulocyte Colony-Stimulating Factor on Neutrophil Apoptosis, Receptor Expression, and Bactericidal Function. Proc Assoc Am Phys 108:455–466

Sullivan R, Fredette JP, Leavitt JL, Gadenne AS, Griffin JD, Simons ER (1989b) Effects of recombinant human granulocyte-macrophage colony-stimulating factor (GM-CSFrh) on transmembrane electrical potentials in granulocytes: relationship between enhancement of ligand-mediated depolarization and augmentation of superoxide anion (O2-) production. J Cell Physiol 139:361–369

Sullivan R, Fredette JP, Socinski M, Elias A, Antman K, Schnipper L, Griffin JD (1989a) Enhancement of superoxide anion release by granulocytes harvested from patients receiving granulocyte-macrophage colony-stimulating factor. Br J Haematol 71:475–479

Sullivan R, Griffin JD, Simons ER, Schafer AI, Meshulam T, Fredette JP, Maas AK, Gadenne AS, Leavitt JL, Melnick DA (1987) Effects of recombinant human granulocyte and macrophage colony-stimulating factors on signal transduction pathways in human granulocytes. J Immunol 139:3422–3430

Sullivan S, Broide DH (1996) Compartmentalization of eosinophil granulocyte-macrophage colony-stimulating factor expression in patients with asthma. J Allergy Clin Immunol 97:966–976

Sung KL, Li Y, Elices M, Gang J, Sriramarao P, Broide DH (1997) Granulocyte-macrophage colony-stimulating factor regulates the functional adhesive state of very late antigen-4 expressed by eosinophils. J Immunol 158:919–927

Suzu S, Yanai N, Saito M, Kawashima T, Kasahara T, Saito M, Takaku F, Motoyoshi K (1990) Enhancement of the antibody-dependent tumoricidal activity of human monocytes by human monocytic colony-stimulating factor. Jpn J Cancer Res 81:79–84

Suzu S, Yokota H, Yamada M, Yanai N, Saito M, Kawashima T, Saito M, Takaku F, Motoyoshi K (1989) Enhancing effect of human monocytic colony-stimulating factor on monocyte tumoricidal activity. Cancer Res 49:5913–5917

Suzuki R, Suzuki S, Igarashi M, Kumagai K (1986) Induction of interleukin 3 but not interleukin 2 or interferon production in the syngeneic mixed lymphocyte reaction. J Immunol 137:1564–1572

Tai PC, Spry CJ (1990) The effects of recombinant granulocyte-macrophage colony-stimulating factor (GM-CSF) and interleukin-3 on the secretory capacity of human blood eosinophils. Clin Exp Immunol 80:426–434

Tai PC, Sun L, Spry CJ (1991) Effects of IL-5, granulocyte/macrophage colony-stimulating factor (GM-CSF) and IL-3 on the survival of human blood eosinophils in vitro. Clin Exp Immunol 85:312–316

Takahashi GW, Andrews DF 3d, Lilly MB, Singer JW, Alderson MR (1993) Effect of granulocyte-macrophage colony-stimulating factor and interleukin-3 on interleukin-8 production by human neutrophils and monocytes. Blood 81:357–364

Takahashi N, Udagawa N, Akatsu T, Tanaka H, Shionome M, Suda T (1991) Role of colony-stimulating factors in osteoclast development. J Bone Miner Res 6:977–985

Tamura M, Hattori K, Nomura H, Oheda M, Kubota N, Imazeki I, Ono M, Ueyama Y, Nagata S, Shirafuji N (1987) Induction of neutrophilic granulocytosis in mice by administration of purified human native granulocyte colony-stimulating factor (G-CSF). Biochem Biophys Res Commun 142:454–460

Tanaka H, Ishikawa K, Nishino M, Shimazu T, Yoshioka T (1996) Changes in granulocyte colony-stimulating factor concentration in patients with trauma and sepsis. J Trauma 40:718–25; discussion 725–726

Tanaka T, Okamura S, Okada K, Suga A, Shimono N, Ohhara N, Hirota Y, Sawae Y, Niho Y (1989) Protective effect of recombinant murine granulocyte-macrophage colony-stimulating factor against Pseudomonas aeruginosa infection in leukocytopenic mice. Infect Immun 57:1792–1799

Tanimoto Y, Takahashi K, Kimura I (1992) Effects of cytokines on human basophil chemotaxis. Clin Exp Allergy 22:1020–1025

Tavernier J, Devos R, Cornelis S, Tuypens T, Van der Heyden J, Fiers W, Plaetinck G (1991) A human high affinity interleukin-5 receptor (IL5R) is composed of an IL5-specific alpha chain and a beta chain shared with the receptor for GM-CSF. Cell 66:1175–1184

Terashima T, Soejima K, Waki Y, Nakamura H, Fujishima S, Suzuki Y, Ishizaka A, Kanazawa M (1995) Neutrophils activated by granulocyte colony-stimulating factor suppress tumor necrosis factor-alpha release from monocytes stimulated by endotoxin. Am J Respir Cell Mol Biol 13:69–73

Terashita M, Kudo C, Yamashita T, Gresser I, Sendo F (1996) Enhancement of delayed-type hypersensitivity to sheep red blood cells in mice by granulocyte colony-stimulating factor administration at the elicitation phase. J Immunol 156:4638–4643

Teshima T, Shibuya T, Harada M, Akashi K, Taniguchi S, Okamura T, Niho Y (1990) Granulocyte-macrophage colony-stimulating factor suppresses induction of neutrophil alkaline phosphatase synthesis by granulocyte colony-stimulating factor. Exp Hematol 18:316–321

Thomassen MJ, Ahmad M, Barna BP, Antal J, Wiedemann HP, Meeker DP, Klein J, Bauer L, Gibson V, Andresen S, et al (1991) Induction of cytokine messenger RNA and secretion in alveolar macrophages and blood monocytes from patients with lung cancer receiving granulocyte-macrophage colony-stimulating factor therapy. Cancer Res 51:857–862

Thomassen MJ, Barna BP, Rankin D, Wiedemann HP, Ahmad M (1989) Differential effect of recombinant granulocyte macrophage colony-stimulating factor on human monocytes and alveolar macrophages. Cancer Res 49:4086–4089

Thomssen H, Kahan M, Londei M (1996) IL-3 in combination with IL-4, induces the expression of functional CD1 molecules on monocytes. Cytokine 8:476–481

Thorens B, Mermod JJ, Vassalli P (1987) Phagocytosis and inflammatory stimuli induce GM-CSF mRNA in macrophages through posttranscriptional regulation. Cell 48:671–679

Thorpe R, Wadhwa M, Bird CR, Mire-Sluis AR (1992) Detection and measurement of cytokines. Blood Rev 6:133–148

Tiegs G, Barsig J, Matiba B, Uhlig S, Wendel A (1994) Potentiation by granulocyte macrophage colony-stimulating factor of lipopolysaccharide toxicity in mice. J Clin Invest 93:2616–2622

Tkatch LS, Tweardy DJ (1993) Human granulocyte colony-stimulating factor (G-CSF), the premier granulopoietin: biology, clinical utility, and receptor structure and function. Lymphokine Cytokine Res 12:477–488

Tobler A, Gasson J, Reichel H, Norman AW, Koeffler HP (1987) Granulocyte-macrophage colony-stimulating factor. Sensitive and receptor-mediated regulation by 1,25-dihydroxyvitamin D3 in normal human peripheral blood lymphocytes. J Clin Invest 79:1700–1705

Toda H, Murata A, Matsuura N, Uda K, Oka Y, Tanaka N, Mori T (1993) Therapeutic efficacy of granulocyte colony stimulating factor against rat cecal ligation and puncture model. Stem Cells (Dayt) 11:228–234

Toda H, Murata A, Oka Y, Uda K, Tanaka N, Ohashi I, Mori T, Matsuura N (1994) Effect of granulocyte-macrophage colony-stimulating factor on sepsis-induced organ injury in rats. Blood 83:2893–2898

Tomioka K, MacGlashan Jr. DW, Lichtenstein LM, Bochner BS, Schleimer RP (1993) GM-CSF regulates human eosinophil responses to F-Met peptide and platelet activating factor. J Immunol 151:4989–4997

Treweeke AT, Aziz KA, Zuzel M (1994) The role of G-CSF in mature neutrophil function is not related to GM-CSF-type cell priming. J Leukoc Biol 55:612–616

Tsuchiya M, Asano S, Kaziro Y, Nagata S (1986) Isolation and characterization of the cDNA for murine granulocyte colony-stimulating factor. Proc Natl Acad Sci USA 83:7633–7637

Uchida K, Yamamoto Y, Klein TW, Friedman H, Yamaguchi H (1992) Granulocyte-colony stimulating factor facilitates the restoration of resistance to opportunistic fungi in leukopenic mice. J Med Vet Mycol 30:293–300

Ulich TR, Busser K, Longmuir KJ (1990a) Cytokine- and calcium ionophore A23187-mediated arachidonic acid metabolism in neutrophils. Cytokine 2:280–286

Ulich TR, del Castillo J, Guo K, Souza L (1989) The hematologic effects of chronic administration of the monokines tumor necrosis factor, interleukin-1, and granulocyte-colony stimulating factor on bone marrow and circulation. Am J Pathol 134:149–159

Ulich TR, del Castillo J, McNiece IK, Yin SM, Irwin B, Busser K, Guo KZ (1990b) Acute and subacute hematologic effects of multi-colony stimulating factor in combination with granulocyte colony-stimulating factor in vivo. Blood 75:48–53

Umeda S, Takahashi K, Shultz LD, Naito M, Takagi K (1996) Effects of macrophage colony-stimulating factor on macrophages and their related cell populations in the osteopetrosis mouse defective in production of functional macrophage colony-stimulating factor protein. Am J Pathol 149:559–574

Vadas MA, Nicola NA, Metcalf D (1983) Activation of antibody-dependent cell-mediated cytotoxicity of human neutrophils and eosinophils by separate colony-stimulating factors. J Immunol 130:795–799

Valent P, Besemer J, Muhm M, Majdic O, Lechner K, Bettelheim P (1989a) Interleukin 3 activates human blood basophils via high-affinity binding sites. Proc Natl Acad Sci USA 86:5542–5546

Valent P, Schmidt G, Besemer J, Mayer P, Zenke G, Liehl E, Hinterberger W, Lechner K, Maurer D, Bettelheim P (1989b) Interleukin-3 is a differentiation factor for human basophils. Blood 73:1763–1769

Valerius T, Repp R, de Wit TP, Berthold S, Platzer E, Kalden JR, Gramatzki M, van de Winkel JG (1993) Involvement of the high-affinity receptor for IgG (Fc gamma RI; CD64) in enhanced tumor cell cytotoxicity of neutrophils during granulocyte colony-stimulating factor therapy. Blood 82:931–939

van der Wouw PA, van Leeuwen R, van Oers RH, Lange JM, Danner SA (1991) Effects of recombinant human granulocyte colony-stimulating factor on leucopenia in zidovudine-treated patients with AIDS and AIDS related complex, a phase I/II study. Br J Haematol 78:319–324

van Pelt LJ, Huisman MV, Weening RS, von dem Borne AE, Roos D, van Oers RH (1996) A single dose of granulocyte-macrophage colony-stimulating factor induces systemic interleukin-8 release and neutrophil activation in healthy volunteers. Blood 87:5305–5313

Vecchiarelli A, Monari C, Baldelli F, Pietrella D, Retini C, Tascini C, Francisci D, Bistoni F (1995) Beneficial effect of recombinant human granulocyte colony-stimulating factor on fungicidal activity of polymorphonuclear leukocytes from patients with AIDS. J Infect Dis 171:1448–1454

Vellenga E, Dokter W, de Wolf JT, van de Vinne B, Esselink MT, Halie MR (1991) Interleukin-4 prevents the induction of G-CSF mRNA in human adherent monocytes in response to endotoxin and IL-1 stimulation. Br J Haematol 79:22–26

Vellenga E, Rambaldi A, Ernst TJ, Ostapovicz D, Griffin JD (1988) Independent regulation of M-CSF and G-CSF gene expression in human monocytes. Blood 71:1529–1532

Vial T, Descotes J (1995) Clinical toxicity of cytokines used as haemopoietic growth factors. Drug Saf 13:371–406

Villalta F, Kierszenbaum F (1986) Effects of human colony-stimulating factor on the uptake and destruction of a pathogenic parasite (Trypanosoma cruzi) by human neutrophils. J Immunol 137:1703–1707

Vollmar B, Messner S, Wanner G, Hartung T, Menger MD (1997) Immunomodulatory action of G-CSF in a rat model of endotoxin-induced liver injury: an intravital microscopic analysis of Kupffer cell and leukocyte response. J Leuk Biol 62:710–718

von Ruden T, Mouchiroud G, Bourette RP, Ouazana R, Blanchet JP, Wagner EF (1991) Expression of human CSF-1 receptor induces CSF-1-dependent proliferation in murine myeloid but not in T-lymphoid cells. Leukemia 5:3–7

Wang JM, Chen ZG, Colella S, Bonilla MA, Welte K, Bordignon C, Mantovani A (1988) Chemotactic activity of recombinant human granulocyte colony-stimulating factor. Blood 72:1456–1460

Wang JM, Colella S, Allavena P, Mantovani A (1987) Chemotactic activity of human recombinant granulocyte-macrophage colony-stimulating factor. Immunology 60:439–444

Wang M, Friedman H, Djeu JY (1989) Enhancement of human monocyte function against Candida albicans by the colony-stimulating factors (CSF): IL-3, granulocyte-macrophage-CSF, and macrophage-CSF. J Immunol 143:671–677

Warren MK, Ralph P (1986) Macrophage growth factor CSF-1 stimulates human monocyte production of interferon, tumor necrosis factor, and colony stimulating activity. J Immunol 137:2281–2285

Warringa RA, Koenderman L, Kok PT, Kreukniet J, Bruijnzeel PL (1991) Modulation and induction of eosinophil chemotaxis by granulocyte-macrophage colony-stimulating factor and interleukin-3. Blood 77:2694–2700

Watari K, Asano S, Shirafuji N, Kodo H, Ozawa K, Takaku F, Kamachi S (1989) Serum granulocyte colony-stimulating factor levels in healthy volunteers and patients with various disorders as estimated by enzyme immunoassay. Blood 73:117–122

Weisbart RH, Gasson JC, Golde DW (1989) Colony-stimulating factors and host defense. Ann Intern Med 110:297–303

Weisbart RH, Golde DW, Clark SC, Wong GG, Gasson JC (1985) Human granulocyte-macrophage colony-stimulating factor is a neutrophil activator. Nature 314:361–363

Weisbart RH, Golde DW, Gasson JC (1986) Biosynthetic human GM-CSF modulates the number and affinity of neutrophil f-Met-Leu-Phe receptors. J Immunol 137:3584–3587

Weisbart RH, Kacena A, Schuh A, Golde DW (1988) GM-CSF induces human neutrophil IgA-mediated phagocytosis by an IgA Fc receptor activation mechanism. Nature 332:647–648

Weisbart RH, Kwan L, Golde DW, Gasson JC (1987) Human GM-CSF primes neutrophils for enhanced oxidative metabolism in response to the major physiological chemoattractants. Blood 69:18–21

Weiser WY, Van Niel A, Clark SC, David JR, Remold HG (1987) Recombinant human granulocyte/macrophage colony-stimulating factor activates intracellular killing of Leishmania donovani by human monocyte-derived macrophages. J Exp Med 166:1436–1446

Weiss M, Gross-Weege W, Schneider M, Neidhardt H, Liebert S, Mirow N, Wernet P (1995) Enhancement of neutrophil function by in vivo filgrastim treatment for prophylaxis of sepsis in surgical intensive care patients. J Crit Care 10:21–26

Weiss M, Gross-Weege W, Wernet P (1992) Rapid neutrophil recovery from acquired agranulocytosis by recombinant human granulocyte-macrophage colony-stimulating factor in an intensive care patient. Crit Care Med 20:1490–1491

Weiss M, Groß-Weege W, Wernet P, Neidhardt H, Harms B, Schneider M (1994) Filgrastim (rhG-CSF) for prophylaxis and therapy of sepsis in surgical intensive care patients: Neutrophil function. Clin Intens Care, Suppl 5:96

Welte K, Bonilla MA, Gillio AP, Boone TC, Potter GK, Gabrilove JL, Moore MA, O'Reilly RJ, Souza LM (1987) Recombinant human granulocyte colony-stimulating factor. Effects on hematopoiesis in normal and cyclophosphamide-treated primates. J Exp Med 165:941–948

Welte K, Dale D (1996) Pathophysiology and treatment of severe chronic neutropenia. Ann Hematol 72:158–165

Wheeler JG, Givner LB (1992) Therapeutic use of recombinant human granulocyte-macrophage colony-stimulating factor in neonatal rats with type III group B streptococcal sepsis. J Infect Dis 165:938–941

Wheeler JG, Huffine ME, Childress S, Sikes J (1994) Comparison of colony stimulation factors on in vitro rat and human neutrophil function. Biol Neonate 66:214–20

Whetton AD (1990) The biology and clinical potential of growth factors that regulate myeloid cell production. Trends Pharmacol Sci 11:285–289

Wieser M, Bonifer R, Oster W, Lindemann A, Mertelsmann R, Herrmann F (1989) Interleukin-4 induces secretion of CSF for granulocytes and CSF for macrophages by peripheral blood monocytes. Blood 73:1105–1108

Wiktor-Jedrzejczak W, Urbanowska E, Aukerman SL, Pollard JW, Stanley ER, Ralph P, Ansari AA, Sell KW, Szperl M (1991) Correction by CSF-1 of defects in the osteopetrotic op/op mouse suggests local, developmental, and humoral requirements for this growth factor. Exp Hematol 19:1049–1054

Williams GT, Smith CA, Spooncer E, Dexter TM, Taylor DR (1990) Haemopoietic colony stimulating factors promote cell survival by suppressing apoptosis. Nature 343:76–79

Williams MA, Kelsey SM, Collins PW, Gutteridge CN, Newland AC (1995) Administration of rHuGM-CSF activates monocyte reactive oxygen species secretion and adhesion molecule expression in vivo in patients following high-dose chemotherapy. Br J Haematol 90:31–40

Willman CL, Stewart CC, Miller V, Yi TL, Tomasi TB (1989) Regulation of MHC class II gene expression in macrophages by hematopoietic colony-stimulating factors (CSF). Induction by granulocyte/macrophage CSF and inhibition by CSF-1. J Exp Med 170:1559–1567

Wiltschke C, Krainer M, Wagner A, Linkesch W, Zielinski CC (1995) Influence of in vivo administration of GM-CSF and G-CSF on monocyte cytotoxicity. Exp Hematol 23:402–406

Wodnar-Filipowicz A, Heusser CH, Moroni C (1989) Production of the haemopoietic growth factors GM-CSF and interleukin-3 by mast cells in response to IgE receptor-mediated activation. Nature 339:150–152

Wong GG, Witek JS, Temple PA, Wilkens KM, Leary AC, Luxenberg DP, Jones SS, Brown EL, Kay RM, Orr EC, et al (1985) Human GM-CSF: molecular cloning of the complementary DNA and purification of the natural and recombinant proteins. Science 228:810–815

Xing Z, Ohkawara Y, Jordana M, Graham F, Gauldie J (1996) Transfer of granulocyte-macrophage colony-stimulating factor gene to rat lung induces eosinophilia, monocytosis, and fibrotic reactions. J Clin Invest 97:1102–1110

Yamaguchi M, Hirai K, Morita Y, Takaishi T, Ohta K, Suzuki S, Motoyoshi K, Kawanami O, Ito K (1992b) Hemopoietic growth factors regulate the survival of human basophils in vitro. Int Arch Allergy Immunol 97:322–329

Yamaguchi M, Hirai K, Shoji S, Takaishi T, Ohta K, Morita Y, Suzuki S, Ito K (1992a) Haemopoietic growth factors induce human basophil migration in vitro. Clin Exp Allergy 22:379–383

Yamamoto Y, Klein TW, Friedman H, Kimura S, Yamaguchi H (1993) Granulocyte colony-stimulating factor potentiates anti-Candida albicans growth inhibitory activity of polymorphonuclear cells. FEMS Immunol Med Microbiol 7:15–22

Yang YC, Ciarletta AB, Temple PA, Chung MP, Kovacic S, Witek-Giannotti JS, Leary AC, Kriz R, Donahue RE, Wong GG, et al (1986) Human IL-3 (multi-CSF): identification by expression cloning of a novel hematopoietic growth factor related to murine IL-3. Cell 47:3–10

Yang YC, Kovacic S, Kriz R, Wolf S, Clark SC, Wellems TE, Nienhuis A, Epstein N (1988) The human genes for GM-CSF and IL 3 are closely linked in tandem on chromosome 5. Blood 71:958–961

Yasuda H, Ajiki Y, Shimozato T, Kasahara M, Kawada H, Iwata M, Shimizu K (1990) Therapeutic efficacy of granulocyte colony-stimulating factor alone and in combination with antibiotics against Pseudomonas aeruginosa infections in mice. Infect Immun 58:2502–2509

Yong KL (1996) Granulocyte colony-stimulating factor (G-CSF) increases neutrophil migration across vascular endothelium independent of an effect on adhesion: comparison with granulocyte-macrophage colony-stimulating factor (GM-CSF). Br J Haematol 94:40–47

Yong KL, Linch DC (1992) Differential effects of granulocyte- and granulocyte-macrophage colony-stimulating factors (G- and GM-CSF) on neutrophil adhesion in vitro and in vivo. Eur J Haematol 49:251–259

Yong KL, Linch DC (1993) Granulocyte-macrophage-colony-stimulating factor differentially regulates neutrophil migration across IL-1-activated and nonactivated human endothelium. J Immunol 150:2449–2456

Yuo A, Kitagawa S, Motoyoshi K, Azuma E, Saito M, Takaku F (1992) Rapid priming of human monocytes by human hematopoietic growth factors: granulocyte-macrophage colony-stimulating factor (CSF), macrophage-CSF, and interleukin-3 selectively enhance superoxide release triggered by receptor-mediated agonists. Blood 79:1553–1557

Yuo A, Kitagawa S, Ohsaka A, Ohta M, Miyazono K, Okabe T, Urabe A, Saito M, Takaku F (1989) Recombinant human granulocyte colony-stimulating factor as an activator of human granulocytes: potentiation of responses triggered by receptor-mediated agonists and stimulation of C3bi receptor expression and adherence. Blood 74:2144–2149

Yuo A, Kitagawa S, Ohsaka A, Saito M, Takaku F (1990) Stimulation and priming of human neutrophils by granulocyte colony-stimulating factor and granulocyte-macrophage colony-stimulating factor: qualitative and quantitative differences. Biochem Biophys Res Commun 171:491–497

Yurdakok M, Gurakan B, Ergin H, Kirazli S (1994) G-CSF and GM-CSF in respiratory distress syndrome [letter]. Am J Hematol 46:155–156

Zeck-Kapp G, Kapp A, Riede UN (1989) Activation of human polymorphonuclear neutrophilic granulocytes by immuno-modulating cytokines: an ultrastructural study. Immunobiology 179:44–55

Zsebo KM, Yuschenkoff VN, Schiffer S, Chang D, McCall E, Dinarello CA, Brown MA, Altrock B, Bagby Jr. GC (1988) Vascular endothelial cells and granulopoiesis: interleukin-1 stimulates release of G-CSF and GM-CSF. Blood 71:99–103

Zuckerman SH, O'Neal L (1994) Endotoxin and GM-CSF-mediated down-regulation of macrophage apo E secretion is inhibited by a TNF-specific monoclonal antibody. J Leukoc Biol 55:743–748

List of Abbreviations

ADCC	antibody dependent cellular cytotoxicity
ALL	acute lymphoblastic leukemia
AML	acute myelogenous leukemia
AMP	adenosine monophosphate
ANC	absolute neutrophil count
ARDS	adult respiratory distress syndrome
AUL	acute undifferentiated leukemia
BPI	bacterial permeability increasing protein
CAM	cellular adhesion molecule
CD11b/CD18	Mac-1; C3bi
CD14	LPS-binding protein receptor
CD16	FcγRIII; IgG receptor
CD18	β_2 integrin
CD23	FcεRII; IgE receptor type
CD25	IL-2 receptor
CD32	FcγRII; IgG receptor
CD35	CR-1
CD50	ICAM-3
CD54	ICAM-1
CD62L	LAM-1; L-selectin
CD64	FcγRI; IgG receptor
CD89	IgA receptor
cDNA	complementary DNA
CFU	colony-forming unit
ConA	Concanavalin A
db-cAMP	2'-O-dibutyryl-cAMP
ECP	eosinophil cationic protein
EGF	epidermal growth factor
fMLP	N-formylmethionyl-leucyl-phenylalanine
G-CSF	granulocyte colony-stimulating factor
G-CSF-R	granulocyte colony-stimulating factor receptor
GM-CSF	granulocyte-macrophage colony-stimulating factor
GM-CSF-R	granulocyte macrophage colony-stimulating factor receptor
GVHD	graft-versus-host-disease
HIV	human immunodeficiency virus
i.p.	intraperitoneal
i.v.	intravenous
IFN	interferon

IGF-1	insulin-like growth factor-1
IL	interleukin
IL-1ra	interleukin-1 receptor antagonist
IL-3	interleukin-3 = multi-CSF
IL-3-R	interleukin-3 receptor
LAK	lymphokine-activated killer
LPS	lipopolysaccharide; endotoxin
LT	lymphotoxin
LTA	lipoteichoic acid
LTB_4	leukotriene B_4
M-CSF	macrophage colony-stimulating factor
MPO	myeloperoxidase
multi-CSF	multi-colony-stimulating factor = IL-3
NAP	neutrophil alkaline phosphatase
NK	natural killer cell
PAF	platelet-activating factor
PAI	plasminogen-activator inhibitor
PBMC	peripheral blood mononuclear cells
PFC	plaque-forming cell
PGE_2	prostaglandin E_2
PHA	phytohemagglutinin
PMA	phorbol myristate acetate
PMN	polymorphonuclear granulocyte = neutrophil
rhu	recombinant human
rmu	recombinant murine
ROS	reactive oxygen species
s.c.	subcutaneous
SCN	severe congenital neutropenia = Kostmann Syndrome
SEA	staphylococcal enterotoxin A
SEB	staphylococcal enterotoxin B
sTNF-R	soluble TNF receptor
TGF-β	transforming growth factor β
TNF	tumor necrosis factor
TRAP	tartrate-resistent acid phosphatase
u-PA	urokinase-type plasminogen activator
VLA-4	very late antigen 4
WBC	white blood cell count

Editor-in-charge: Professor D. Pette

Hyperpolarization-Activated Cation Channels: A Multi-Gene Family

M. Biel, A. Ludwig, X. Zong and F. Hofmann

Institut für Pharmakologie und Toxikologie der Technischen Universität München, Biedersteiner Straße 29, D-80802 München, Germany

Contents

1 Introduction

A variety of body functions like heart beat, sleep-wake cycle, secretion of hormones and control of behavioural state, depend on the action of so-called pacemakers, specialized cells that are able to generate rhythmic, spontaneously firing action potentials. The archetypal organ displaying autonomic rhythmicity is the heart. Pacemaking of the heart is accomplished by the rhythmic discharge of the sinoatrial node (DiFrancesco 1993, 1996; Brown and Ho 1996). The firing rate of the sinoatrial node cells is determined by the diastolic depolarization phase of the action potential. During this phase the membrane potential is slowly depolarized to the threshold triggering the next action potential. The ionic conductance underlying the cardiac pacemaker depolarization was identified in the late seventies and early eighties (Brown et al. 1977; Yanagihara and Irisawa 1980; DiFrancesco 1981) and called I_f (f for "funny") or I_h (h for hyperpolarization activated). At about the same time a similar current was discovered in neurons, first in photoreceptors (Attwell and Wilson 1980; Bader and Bertrand 1984; Barnes and Hille 1989) and then in various central neurons e.g. hippocampal pyramidal cells (Halliwell and Adams 1982) where it was called I_q (q for"queer"). Later on this type of current was found in a rich diversity of central and peripheral neurons (Pape 1996). The common properties of $I_h/I_f/I_q$ are (1) activation by hyperpolarization negative to potentials of –50 to –70 mV; (2) conductance of Na^+ and K^+ ions and (3) enhancement by cyclic AMP. Intriguingly, cAMP increases I_h by a mechanism that is independent of protein phosphorylation, but which involves direct binding of the cyclic nucleotide to the channel that mediates I_h (DiFrancesco and Tortora 1991; Pape and McCormick 1989). These unique properties make the current ideally suited for generating the depolarization of cardiac pacemaker cells (DiFrancesco 1993, 1996; Brown and Ho 1996). Hyperpolarization at the termination of the cardiac action potential activates I_h which then gradually depolarizes the membrane. Acceleration of the heart rate in response to sympathetic stimulation is due to activation of β adrenergic receptors leading to activation of G_s protein and in turn adenylyl cyclase. The resulting increase in intracellular cAMP levels directly activates the I_h channel. cAMP binding to the channel leads to a shift of the activation curve towards more positive voltages. This shift results in an increased inward current at a fixed membrane potential and therefore an acceleration of the diastolic depolarization. Muscarinic stimulation leads to an opposite effect, in part due to a decrease in cAMP levels and reduction of the I_h current (DiFrancesco er al. 1989; see also Wickman et al. 1998).

In the central nervous system, the role of I_h is likely to be more complex than in heart (for detailed reviews see Pape 1996; Lüthi and McCormick 1998). Like its cardiac counterpart, the neuronal I_h current critically controls the rate of rhythmic oscillations of single neurons and neuronal networks ("neuronal pacemaking"). Additionally, at least two other roles for neuronal I_h have been demonstrated: determination of the resting membrane potential and contribution to the neuronal response to hyperpolarizing currents.

Due to the lack of cloned cDNAs, I_h channels could only be investigated in native cells up to now. The recent molecular identification of five different genes (Gauss et al. 1998; Ludwig et al. 1998; Santoro et al. 1998) encoding I_h channels has now provided the tools for a detailed analysis of molecular mechanisms underlying I_h channel function. In this review we will mainly focus on the discussion of the structure and function of the cloned channels and will, whenever feasible, correlate the properties of cloned and native channels.

2 Molecular Cloning of Hyperpolarization-Activated Cation Channels

Although several attempts were made in the last ten years to clone the genes underlying I_h the cDNAs of this channel class were identified only recently. A first clone was found in mouse brain (BCNG-1), via a yeast two-hybrid screen designed to detect proteins interacting with N-Src (Santoro et al. 1997). However, since the clone could not be functionally expressed in the initial study, assignment as a cyclic nucleotide-modulated K^+ channel was based on sequence alignments to cyclic nucleotide-gated (CNG) and K^+ channels. Later on it turned out that BCNG-1 was indeed a member of an extended family of hyperpolarization-activated cation channels (Santoro et al. 1998). Independently of Santoro et al., channels underlying I_h were cloned based on the idea that I_h channels like CNG channels (Zagotta and Siegelbaum 1996; Finn et al. 1996; Biel et al. 1998) or cyclic nucleotide-dependent protein kinases (Pfeifer et al. in press) should contain a cyclic nucleotide-binding domain (CNBD). By screening the EST data base for sequences matching the CNBD of CNG channels, a partial sequence was identified and subsequently used to isolate three homologous cDNAs from mouse brain (Ludwig et al. 1998). Similarly, a PCR-based approach using degenerate primers to amplify sequences of new CNBDs resulted in the isolation of an I_h channel from sea urchin testis (Gauss et al. 1998). Three different names have been proposed for the clones underlying I_h. To avoid misunderstandings arising from the use of different designations for the same protein, it is

Table 1. Functional properties of cloned hyperpolarization-activated cation channels

old name proposed new name	HAC2[1], mBCNG-1[2], mHCN1	HAC1[1], mBCNG-2[2] mHCN2	HAC3[1], mBCNG-4[2] mHCN3	mBCNG-3[2] mHCN4	SPIH[3] spHCN
species	mouse, human	mouse, human	mouse	mouse	sea urchin
amino acids	910	863	779	partial sequence	767
tissue distribution	brain	brain, heart	brain	brain, heart	sperm
τ activation	98 ms at −130 mV	241 ms at −140 mV	n.d	n.d.	n.d
$V_{½}$[4)] ~ 5 mM external K^+	−100 mV	−100 mV	n.d	n.d.	−26 mV
P_{Na}/P_K	0.25	0.24	n.d	n.d.	0.23
K_a for cAMP/Hill coefficient	n.d.	0.5 μM/0.8	n.d.	n.d.	0.74 μM/1.05
shift by cAMP	+1.8 mV	+13 mV	n.d.	n.d.	−24.7 mV
inrease of I_{max} induced by cAMP	no	no	n.d.	n.d.	20 fold
K_a for cGMP/Hill coefficient	n.d.	6 μM/1.2	n.d.	n.d.	not sensitive

[1)] Ludwig et al. 1998, [2)] Santoro et al. 1998; [3)] Gauss et al. 1998, [4)] measured in whole-cell voltage clamp mode.

clearly desirable to find a common nomenclature for the channels. As has been recently proposed (Clapham 1998) we suggest HCN for Hyperpolarization-activated, Cyclic Nucleotide-gated channel as an adequate new name. Table 1 summarizes old and new designations for the different members of the channel family.

3 Mammalian HCN Channels

Four different members of the HCN family (HCN1-4) have been identified in mouse and human (Fig. 1). The channels are closely related to each other

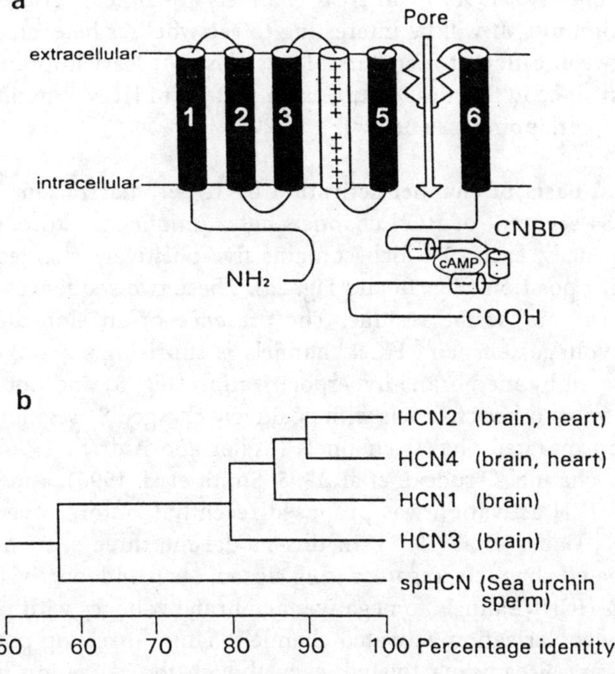

Fig. 1. **a** Model of the transmembrane structure of HCN channels. The six transmembrane segments S1-S6 are numbered 1–6, the regularly spaced positively charged residues in the voltage-sensing S4 segment are indicated by plus signs. CNBD, cyclic nucleotide-binding domain. The arrow illustrates inward cation flux through the pore region. **b** Phylogenetic tree of the HCN channel family. The tree was calculated by comparison of the corresponding regions from segment S1 to the end of the CNBD. The tissue expression of HCN1-4 in mammals and of spHCN in sea urchin is shown to the right of the dendrogramm

having an overall sequence identity of about 60%. The central portion of the channels which includes the transmembrane segments and the CNBD is even higher conserved with a sequence identity of 80–90%, whereas N- and C-termini diverge more strongly between the different genes. Hydropathicity analysis and sequence alignments indicate that HCN channels belong to the superfamily of voltage-gated cation channels. Like K^+-channels (Jan and Jan 1994; MacKinnon 1995, Pongs 1996), they contain six transmembrane helices (S1–S6) including a positively charged S4 segment, and an ion-conducting pore loop between the fifth and sixth transmembrane helix. Similar to CNG channels, HCN channels contain a CNBD in the cytosolic C-terminus which is separated from the last transmembrane helix by a linker region. Thus, HCN channels combine structural features common to both voltage-gated K^+-channels and CNG channels. Like these channels (Doyle et al. 1998; Liu et al. 1996; 1998) HCN channels are almost certainly composed of four subunits. It will be interesting to see whether heteromeric channels exist between different members of this class. At least in brain heteromers may exist since in this tissue expression pattern of HCN subunits are at least partially overlapping (see below).

Structural basis of channel activation by hyperpolarization. The voltage-sensing S4 segment of HCN channels has a unique structure, consisting of two sequences, each of which contains five positively charged residues at every third position (Fig. 1a and Fig. 2a). These two sequences are separted by an 'in-frame' serine residue. The presence of an elongated positively charged voltage-sensor in HCN channels is suprising since these channels are activated by membrane hyperpolarization (Fig. 3) and not by depolarization as most other channels with positively charged S4 segments. Based on studies on mutated Shaker channels (Miller and Aldrich 1996) and on the HERG K^+ channel (Trudeau et al. 1995; Smith et al. 1996), a mechanism explaining HCN activation was proposed recently (Santoro et al. 1998; Clapham 1998; Gauss et al. 1998). In this model the three sequential states of voltage-gated channels, respresenting closed, open and inactivated pores are shifted in HCN channels to negative membrane voltages with respect to the states of depolarization-activated channels. Thus, at resting potential, HCN channels are already inactivated, even though the activation gates (the S4 segments) may be in the open configuration. Hyperpolarization would then open the channels simply by reversing the inactivation reaction. The molecular structure of the "inactivation gate" as well as the mechanism coupling S4 movement to pore gating is not known at present. However, a first clue towards an understanding of HCN activation is provided by a recent study on the inwardly rectifying K^+ channel KAT1 from *Arabidopsis thaliana*

(Marten and Hoshi 1998). KAT1 resembles HCN channels in that it is both activated by hyperpolarization and also contains a positively charged S4 segment (Fig. 2a). By investigating the effect of N-terminal deletions and a S4 mutation in the KAT1 channel Marten and Hoshi concluded that in this channel the hyperpolarizing shift in the activation curve is due to the interaction of cytoplasmic N-terminus with the S4 segment. A similar mechanism may also exist in HCN channels.

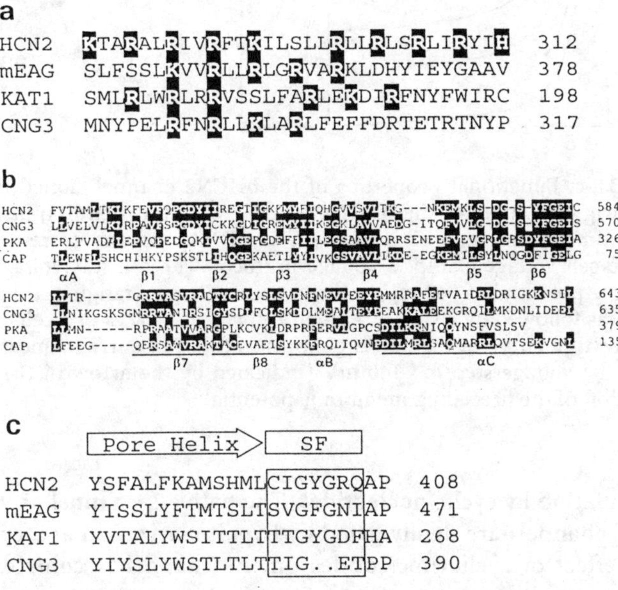

Fig. 2. **a** Multiple sequence alignment of the putative voltage-sensing S4 segment of HCN2 with that of other channels. Basic residues at every third position are highlighted in black with reverse type. mEAG , mouse ether-à-gogo potassium channel (Warmke and Ganetzky 1994); KAT1, hyperpolarization-activated potassium channel of *Arabidopsis thaliana* (Anderson et al. 1992); CNG3, bovine cone photoreceptor channel (Biel et al. 1994). **b** Comparison of the CNBDs from various proteins. PKA, bovine cAMP-dependent protein kinase (Titani et al. 1984); CAP, catabolite activator protein of *E. coli* (Weber and Steitz 1987). α(A–C) helices and ß (1–8) strands in the CAP crystal structure are underlined. Residues identical in at least two sequences are highlighted by a black background. The residue D604 controling ligand selectivity in rod photoreceptor CNG channel (see text) is equivalent to D628 of CNG3 and I636 in HCN2. **c** Comparison of the pore region of HCN with that of other channels. The pore helix and selectivity filter (SF) are indicated according to the crystal structure of the *Streptomyces lividans* potassium channel (Doyle et al. 1998)

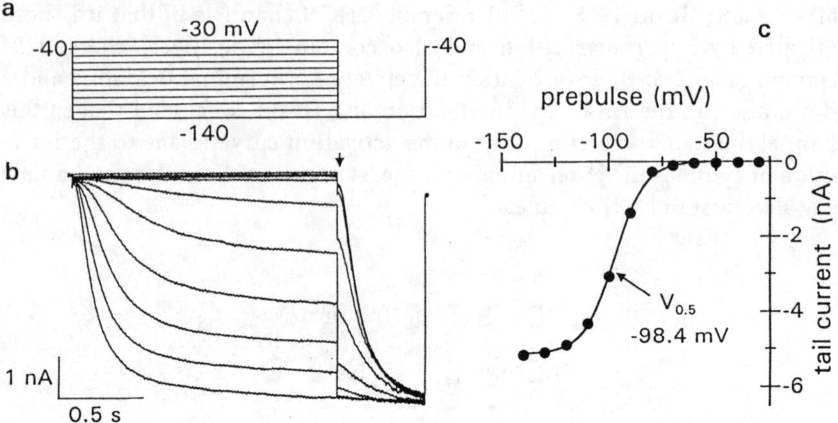

Fig. 3a–c. Functional properties of the hHCN2 channel cloned from human heart (Ludwig et al. unpublished). HEK293 cells were transiently transfected with an expression vector containing the hHCN2 cDNA and the current was measured in whole-cell voltage clamp. **a** voltage protocol. The cell was voltage clamped from a holding potential of −40 mV to the various voltages (ranging from −140 to −30 mV); this was followed by a step to −140 mV. **b** current traces of a cell expressing hHCN2. **c** activation curve of the current shown in **(b).** Tail currents measured immediately after the voltage step to −140 mV (indicated by the arrow in **(b)** were plotted as a function of the preceding membrane potential

Modulation by cyclic nucleotides. The native I_h channel as well as expressed HCN channels are dually gated by hyperpolarization and cyclic nucleotides. The effect of cyclic nucleotides on the channel is complex comprising at least two components, a 2–13 mV shift of the activation curve towards positive membrane potentials and an acceleration of the activation kinetics. Cyclic nucleotides regulate HCN channel activity by directly binding to a CNBD which is situated in the C-terminus of the protein. The CNBD of HCN channels reveals a striking sequence similarity with CNBDs of other cyclic nucleotide-regulated proteins like the catabolite activator protein (CAP) of *E. coli*, cAMP- and cGMP-dependent protein kinases and CNG channels (Fig. 2b). Amino acids which have been determined in the CAP crystal structure to lie close to the cAMP molecule (Weber and Steitz 1987) are well conserved in HCN channels. The native I_h channel (DiFrancesco and Tortora 1991) and also the expressed mHCN2 channel reveal an about 10 fold higher apparent affinity for cAMP than for cGMP (see Table 1) whereas CNG channels are about 40–100 fold more sensitive for cGMP than for cAMP (Zagotta and Siegelbaum 1996). Mutagenesis studies indicated that a negatively charged amino acid in the αC helix of the CNBD (D604 in the rod photoreceptor channel; see legend to Fig. 2b) is a major determinant of selectivity for cGMP in CNG channels (Varnum et al. 1995). Replacement of D604 by neutral amino acids results in CNG channels that no longer select cGMP over cAMP. The amino acid corresponding to D604 is replaced by an isoleucine residue in HCN channels being consistent with the notion that the αC helix also critically controls ligand selectivity in these channels. In contrast, a threonine residue in the β7 roll which determines cGMP selectivity in cyclic nucleotide-dependent protein kinases (Shabb et al. 1990) is unlikely to be involved in determining ligand selectivity in cyclic nucleotide-dependent cation channels since it is present in both cAMP-selective HCN channels (T592 in mHCN2) and cGMP-selective CNG channels (T584 in CNG3). The activation of native (DiFrancesco and Tortora 1991) and heterologously expressed HCN channels by cAMP reveals no cooperativity (Hill coefficients of about 1; see Table 1). This finding is very surprising since the activation by cyclic nucleotides of the structurally related CNG channel is a highly cooperative process. Thus, despite the high degree of sequence similarity in the CNBDs of various cyclic nucleotide-binding proteins, the mechanism coupling cyclic nucleotide-binding to channel activation may be significantly different in each case.

Ion selectivity. Under physiological conditions, the I_h current is carried by both Na^+ and K^+. The native I_h channel slightly (about 4 fold) selects K^+ over Na^+, whereas it is almost impermeable to Li^+ (Wollmuth and Hille 1992; Ho et al. 1994). Divalents and anions cannot pass through the I_h channel (Frace et al. 1992a; Pape 1996). The ion selectivity of the heterologously expressed HCN channels concurs well with the relative permeation ratios of native channels. Sequence analysis of cloned HCN channels revealed that the putative ion conducting pore region of these channels is much more related to the pore of K^+-selective channels (Doyle et al. 1998; Choe et al. 1998) than one might have been expected. The pore of K^+ channels is formed by a short loop structure localized between the fifth and sixth putative transmembrane segment (Fig. 1a and 2c). Three amino acids, glycine-tyrosine-glycine, in the centre of the pore loop are a hallmark of almost all K^+ pores (in some channels like the ether-à-gogo (EAG) K^+-channel, the tyrosine residue is replaced by phenylalanine, see Fig. 2c). The GYG sequence has been shown to form the selectivity filter of K^+ channels (Heginbotham et al. 1994; Doyle et al. 1998). Surprisingly, HCN channels also contain a GYG sequence, although they pass both Na^+ and K^+. Thus, a GYG sequence alone is not sufficient to mediate selectivity for K^+. The recent determination of the crystal structure of the bacterial KcsA K^+ channel (Doyle et al. 1998) gives a first clue to the question why HCN channels may have lost K^+ selectivity. In K^+-selective channels the narrowest part of the pore is formed by the carbonyl backbones of the GYG residues which are constrained in an optimal geometry so that only α K^+ ion fits with proper coordination. The strutural rigidity of the carbonyl backbone is provided by two layers of aromatic amino acids that form a cuff around the pore, presumably pulling the pore opening to the correct distance like a ring of springs. In HCN channels the carbonyl backbone simply may have lost some of its structural rigidity, thus allowing both K^+ and Na^+ to pass the pore. The pore of HCN channels carries several amino acids that are not conserved with respect to the pore of selective K^+ channels (Fig. 2c). These amino acid replacements may provide the structural basis that underlies the "relaxation" of ion selectivitiy in HCN channels.

Potassium is not only a permeating ion of HCN channels, it also regulates permeation of Na^+ in a complex fashion. Both the current amplitude and the P_{Na}/P_K ratio of HCN channels depend on the extracellular K^+ concentration i.e. an increase in extracellular K^+ concentration results in a strongly increased current amplitude and in a slightly reduced selectivity for K^+ over Na^+ (Frace et al. 1992b; Wollmuth and Hille 1992; Brown and Ho 1996). The striking interdependence of Na^+ and K^+ permeation in HCN channels is illustrated by the finding that the channels conduct little, if any, Na^+ in the absence of K^+ ions. Thus, although Na^+ is the major carrier of inward current in HCN channels at a range of membrane potentials where the channel is important physiologically, yet K^+ is required for the channel to carry any current. The structural basis of the intimate connection of K^+ and Na^+ permeation pathways is unknown at present. However, it seems reasonable to assume that the pore of HCN is, like that of CNG channels (Sesti et al. 1995) and Ca^{2+} channels (Hess and Tsien 1984), a multi-ion pore possessing at least two cation binding sites: one at the external mouth of the channel having a higher affinity for K^+ and another having a higher affinity for Na^+ (Wollmuth 1995). HCN channels are not only regulated by K^+ but also by external Cl^-. Substitution of external Cl^- by larger anions such as isothionate or gluconate results in a pronounced reduction of the current amplitude of both native (Frace et al. 1992a) and expressed channels (Santoro et al. 1998; Zong et al. unpublished). Thus, the pore of HCN channels is likely to contain a binding site for Cl^- at the extracellular face of the channel. It is tempting to speculate that the positively charged residues present in the pore of HCN channels (Fig. 2c) may be involved in the formation of such an anion-binding domain. The physiological function of the chloride-dependence of HCN channels is not understood yet. It was speculated that Cl^- may perform a screening role for cations bound at external sites of the multi-ion channel and thereby represent a necessary step in channel permeation by cations (Frace et al. 1992a).

4
Correlation Between Native and Cloned HCN Channels

Neuronal HCN channels. Four different HCN channel types have been identified in brain (Table 1). In situ hybridization indicates that HCN2 is the most widely expressed channel type in brain. Transcripts of HCN2 are nearly ubiquitously distributed in brain whereas HCN1 expression is more limited to specific parts of the brain like hippocampal CA1 neurons, superior colliculus, cerebral cortex and cerebellum. The partially overlapping expression pattern of HCN1 and HCN2 suggests that these channels may form heteromers in subsets of neurons. At present, the expression of HCN3 and HCN4 has not been mapped at high resolution in brain, however the rather faint signal observed in Northern blots indicates that at least HCN3 is restricted to specific parts of the brain and/or expressed at much lower density than HCN1 and HCN2. The presence of multiple HCN channels in brain is in good agreement with properties of I_h currents in different types of neurons. The various I_h currents significantly differ in terms of activation kinetics, modulation by cAMP, and voltage-dependence of activation (for review see Pape 1996). The region-specific expression of HCN channels, possibly in combination with the formation of heteromeric channels, could well explain the observed current diversity. This notion is supported by the expression of HCN1 (Santoro et al. 1998) and HCN2 (Ludwig et al. 1998) channels in heterologous expression systems. The expressed channels reveal the principal properties of native I_h channels like activation by hyperpolarization (Fig. 3), permeability for Na^+ and K^+, and blockage by extracellular Cs^+ but also significantly differ from each other in terms of their activation kinetics and modulation by cAMP. HCN1, which is expressed in hippocampal neurons, resembles the native hippocampal channel (Maccaferri et al. 1993; Pedarzani and Storm 1995) in that it activates relatively rapidly upon hyperpolarization and is only weakly shifted by cAMP (1.8 mV). In contrast, HCN2, which is highly expressed in thalamic neurons, resembles channels characterized in these neurons (McCormick and Pape 1990; Pape 1992), in that it activates rather slowly but is profoundly regulated by cAMP (13 mV shift and significant acceleration of activation kinetics).

Cardiac HCN channels. Two HCN channel genes are expressed in mouse and human heart, HCN2 and HCN4. In situ hybridization (Ludwig et al. 1998) and PCR analysis (Santoro et al. 1998) revealed that HCN2 is expressed throughout the heart including the sinoatrial node (Santoro et al. 1998). The exact distribution of HCN4 within the heart has not yet been investigated, however the relatively low abundance of HCN4 clones in human heart cDNA libraries (Ludwig et al. unpublished data) indicates that this channel may exist in much lower concentrations in total heart than HCN2. The presence of HCN channels in a variety of heart cells is consistent with data from the literature describing I_h currents in both spontaneously firing pacemaker cells of sinoatrial node and ventricular myocytes that are normally not involved in generating pacemaker potential (Yu et al. 1995; Baker et al. 1997; Hoppe et al. 1998). The major difference between I_h currents from ventricular myocytes and sinoatrial node cells is refered to the voltage dependence of channel activation. Sinoatrial I_h channels activate at significantly more positive potentials than I_h channels in ventricular myocytes. The expressed HCN2 channel (Ludwig et al. 1998) fits the general properties of sinoatrial channels in terms of activation kinetics, pharmacology and modulation by cyclic nucleotides. However, HCN2 activates at membrane potentials ($V_{½}$ ~ −100 mV) which are more consistent with potentials reported for ventricular (range of $V_{½}$: −95 mV to −135 mV; Yu et al. 1995; Baker et al. 1997; Hoppe et al. 1998) than for sinoatrial node channels (range of $V_{½}$: −65 mV to −90 mV; Denyer and Brown 1990; Tortora and DiFranceso 1991; DiFrancesco and Mangoni 1994). The reason for this discrepancy is unknown at present. An intrinsic factor that is present in sinoatrial node cells but is missing in heterologous expression systems may be necessary to confer channel activation at more positive voltages. Alternatively, the native sinoatrial node channel may be a heteromer consisting of HCN2 and another channel subunit. It is obvious that HCN4 would be a good candidate for this potential second subunit.

5 Invertebrate HCN Channels

The identification of a HCN channel from sea urchin *Strongylocentrotus purpuratus* (original name: SPIH (Gauss et al. 1998); proposed new name: spHCN) indicates that HCN channels have emerged early in evolution. The primary structure of spHCN reveals all the hallmarks of mammalian channels. Especially the core region of the channels including the six transmembrane segments, the pore loop and the CNBD are highly conserved among vertebrate and invertebrate channels (Fig. 1b). This structural similarity concurs well with the variety of functional properties that are shared by both vertebrate and invertebrate channels, namely activation by hyperpolarization, permeation of both Na^+ and K^+, blockage by Cs^+ and sensitivity for cAMP. However, there are also clear differences between the channels (Table 1). For example, spHCN activates at much more positive potentials than mammalian HCN channels. In addition, the spHCN current is transient, whereas mammalian HCN currents reveal no inactivation during maintained hyperpolarization. The most striking difference is refered to the mechanism that underlies cylic nucleotide modulation. cAMP enhances both vertebrate and spHCN channels. However, the augmentation of the current arises in the case of spHCN from an up to 20fold increase in the maximum current whereas in mammalian HCN channels cAMP does not increase maximum current but shifts the activation curve towards positive membrane potentials. In addition, spHCN is not sensitive to cGMP whereas mammalian channels are affected by both cAMP and cGMP. Given the high similarity in the CNBD of spHCN and mammalian channels it will be interesting to see which amino acid residues determine these functional differences.

The functional role of spHCN is only poorly understood at present. The channel is expressed in sperm flagellum and it was postulated that it may be involved in the control of flagellar beating (Gauss et al. 1998). It is unknown at present, whether or not HCN channels are involved in the regulation of sperm motility in mammals. So far, transcripts of HCN1–4 could not be detected in testis (Ludwig et al. 1998; Santoro et al. 1998) indicating that either another channel type is expressed in this tissue or that movement of mammalian sperm does not require the activation of HCN channels.

6
Conclusions

Since the first discovery of an I_h current in heart pacemaker cells about twenty years ago our knowledge on this ion channel class has increased substantially. Studies from a variety of groups have clearly demonstrated that I_h is not only required for cardiac pacemaking but also is a key regulator of several neuronal functions. The long-sought molecular cloning of the genes underlying I_h has now provided the basis to achieve a deeper understanding of the molecular function and physiological regulation of these channels. In addition, the powerful approaches of mouse genetics will enable us to study the physiological roles of the channels in significantly more detail than it was possible up to now.

Acknowledgements. The research conducted in the author's laboratory was supported from Fond der Chemischen Industrie, BMBF and Deutsche Forschungsgemeinschaft.

References

Anderson JA, Huprikar SS, Kochian LV, Lucas WJ, Gaber, RF (1992) Functional expression of a probable *Arabidopsis thaliana* potassium channel in *Saccharomyces cerevisiae*. Proc Natl Acad Sci USA 89:3736–3740

Attwell D, Wilson M (1980) Behaviour of the rod network in the tiger salamander retina mediated by membrane properties of individual rods. J Physiol 309:287–315

Bader CR, Bertrand D (1984) Effect of changes in intra- and extracellular sodium on the inward (anomalous) rectification in salamander photoreceptors. J Physiol 347:611–31

Baker K, Warren KS, Yellen G, Fishman MC (1997) Defective "pacemaker" current (I_h) in a zebrafish mutant with a slow heart rate. Proc Natl Acad Sci 94:4554–4559

Barnes S, Hille B (1989) Ionic channels of the inner segment of tiger salamander cone photoreceptor. J Gen Physiol 94:719–743

Biel M, Sautter A, Ludwig A, Hofmann F, Zong X (1998) Cyclic nucleotide-gated channels – mediators of NO:cGMP-regulated processes. Naunyn Schmiedebergs Arch Pharmacol 358:140–144

Biel M, Zong X, Distler M, Bosse E, Klugbauer N, Murakami M, Flockerzi V, Hofmann F (1994) Another member of the cyclic nucleotide-gated channels family, expressed in testis, kidney, and heart. Proc Natl Sci USA 91:3505–3509

Brown HF, Giles W, Noble SJ (1977) Membrane currents underlying activity in frog sinus venosus. J Physiol 271:783–816

Brown HF, Ho WK (1996) The hyperpolarization-activated inward channel and cardiac pacemaker activity. In: Morad M, Ebashi S, Trautwein W, Kurachi Y (eds) Molecular physiology and pharmacology of cardiac ion channels and transporters. Kluwer Academic Publishers, Dordrecht, pp 17–30

Choe S, Robinson R (1998) An ingenious filter: the structural basis for ion channel selectivity. Neuron 20:821–823

Clapham DE (1998) Not so funny anymore: pacing channels are cloned. Neuron 21:5–7

Denyer JC, Brown HF (1990) Rabbit sino-atrial node cells: isolation and electrophysiological properties. J Physiol 428:405–424

DiFrancesco D (1981) A new interpretation of the pacemaker current in calf Purkinje fibres. J Physiol 314: 359–376

DiFrancesco D (1993) Pacemaker mechanisms in cardiac tissue: Annu Rev Physiol 55: 455–472

DiFrancesco D (1996) The hyperpolarization actiated (i_f) current. Autonomic regulation and the control of pacing. In: Morad M, Ebashi S, Trautwein W, Kurachi Y (eds) Molecular physiology and pharmacology of cardiac ion channels and transporters. Kluwer Academic Publishers, Dordrecht, pp 31–37

DiFrancesco D, Ducouret, Robinson RB (1989) Muscarinic modulation of cardiac rate at low acetylcholine concentrations. Science 243:669–671

DiFrancesco D, Mangoni M (1994) Modulation of single hyperpolaritation-activated channels (i_f) by cAMP in the rabbit sino-atrial node. J Physiol 474:473–482

DiFrancesco D, Tortora P (1991) Direct activation of cardiac pacemaker channels by intracellular AMP. Nature 351:145–147

Doyle DA, Cabral JM, Pfuetzner RA, Kuo A, Gulbis JM, Cohen SL, Chait BT, MacKinnon R (1998) The structure of the potassium channel: molecular basis of K^+ conduction and selectivity. Science 280:69–77

Finn JT, Grunwald ME, Yau KW (1996) Cyclic nucleotide-gated ion channels: an extended family with diverse functions. Annu Rev Physiol 58:395–426

Frace AM, Maruoka F, Noma A (1992a) Control of the hyperpolarization-activated cation current by external anions in rabbit sino-atrial node cells. J Physiol 453:307–318

Frace AM, Maruoka F, Noma A (1992b) External K^+ increases Na^+ conductances of the hyperpolarization-activated current in rabbit cardiac pacemaker cells. Pflügers Arch 421:94–96

Gauss R, Seifert R, Kaupp UB (1998) Molecular identification of a hyperpolarization-activated channel in sea urchin sperm. Nature 393:583–587

Halliwell JV, Adams PR (1982) Voltage-clamp analysis of muscarinic excitation in hippocampal neurons. Brain Res 250:71–92

Heginbotham L, Lu Z, Abramson T, MacKinnon R (1994) Mutations in the K^+ channel signature sequence. Biophys J 66:1061–1067

Hess P, Tsien RW (1984) Mechanism of ion permeation through calcium channels. Nature 309:453–456

Ho WK, Brown HF, Noble D (1994) High selectivity of the i_f channels to Na^+ and K^+ in rabbit isolated sinoatrial node cells. Pflügers Arch 426:68–74

Hoppe UC, Jansen E, Südkamp M, Beuckelmann DJ (1998) Hyperpolarization-activated inward current in ventricular myocytes from normal and failing human hearts. Circulation 97:55–65

Jan LY, Jan YN (1994) Potassium channels and their evolving gates. Nature 371:119–122

Liu DT, Tibbs GR, Paoletti P, Siegelbaum SA (1998) Constraining ligand-binding site stoichiometry suggests that a cyclic nucleotide-gated channel is composed of two functional dimers. Neuron 21:235–248

Liu DT, Tibbs GR, Siegelbaum SA (1996) Subunit stoichiometry of cyclic nucleotide-gated channels and effects of subunit order on channel function. Neuron 16:983–990

Ludwig A, Zong X, Jeglitsch M, Hofmann F, Biel M (1998) A family of hyperpolarization-activated mammalian cation channels. 393:587–591

Lüthi A, McCormick DA (1998) H-Current: Properties of a neuronal and network pacemaker. Neuron 21:9–12

Maccaferri G, Mangoni M, Lazzari A, DiFrancesco D (1993) Properties of the hyperpolarization-activated current in rat hippocampal CA1 pyramidal cells. J Neurophysiol 69:2129–2136

MacKinnon R (1995) Pore loops: an emerging theme in ion channel structure. Neuron 14:889–892

Marten I, Hoshi T (1998) The N-terminus of the K channel KAT1 controls its voltage-dependent gating by altering the membrane electric field. Biophys J 74:2953–2962

McCormick DA, Pape HC (1990) Properties of a hyperpolarization-activated cation current and its role in rhythmic oscillation in thalamic relay neurons. J Physiol 432:291–318

Miller AG, Aldrich RW (1996) Conversion of a delayed rectifier K^+ channel to a voltage-gated inward rectifier K^+ channel by three amino acid substitutions. Neuron 16:853–858

Pape HC (1992) Adenosine promotes burst activity in guinea-pig geniculocortical neurons through two different ionic mechanisms. J Physiol 447:729–753

Pape HC (1996) Queer current and pacemaker: the hyperpolarization-activated cation current in neurons. Annu Rev Physiol 58:299–327

Pape HC, McCormick DA (1989) Noradrenaline and serotonin selectively modulate thalamic burst firing by enhancing a hyperpolarization-activated cation current. Nature 340:715–718

Pedarzani P, Storm JF (1995) Protein kinase A-independent modulation of ion channels in the brain by cyclic AMP. Proc Natl Acad Sci USA 92:11716–11720

Pfeifer A, Dostmann WRG, Sausbier M, Klatt P, Ruth P, Hofmann F. cGMP-dependent protein kinases: structure and function. Rev Physiol Biochem Pharmacol (in press)

Pongs O (1996) Diversity of voltage-dependent K channels. In: Morad M, Ebashi S, Trautwein W, Kurachi Y (eds) Molecular physiology and pharmacology of cardiac ion channels and transporters. Kluwer Academic Publishers, Dordrecht, pp 107–117

Santoro B, Grant SGN, Bartsch D, Kandel ER (1997) Interactive cloning with the SH3 domain of N-src identifies a new brain specific ion channel protein, with homology to Eag and cyclic nucleotide-gated channels. Proc Natl Acad Sci 94:12815–14820

Santoro B, Liu DT, Yao H, Bartsch D, Kandel ER, Siegelbaum SA, Tibbs GR (1998) Identification of a gene encoding a hyperpolarization-activated pacemaker channel of brain. Cell 93:717–729

Sesti F, Eismann E, Kaupp UB, Nizzari M, Torre V (1995) The multi-ion nature of the cGMP-gated channel from vertebrate rods. J Physiol 487:17–36

Shabb JB, Ng L, Corbin JD (1990) One amino acid change produces a high affinity cGMP-binding site in cAMP-dependent protein kinase. J Biol Chem 265:16031–16034

Smith PL, Baukrowitz TM, Yellen G (1996) The inward rectification mechanism of the HERG cardiac potassium channel. Nature 379:833–836

Titani K, Sasagawa T, Ericsson LH, Kumar S, Smith SB, Krebs EG, Walsh KA (1984) Amino acid sequence of the regulatory subunit of bovine type I adenosine cyclic 3',5'-phosphate dependent protein kinase. Biochemistry 23:4193–4199

Trudeau MC, Warmke JW, Ganetzky B, Robertson GA (1995) Herg, a human inward rectifier in the voltage-gated potassium channel family. Science 269:92–95

Varnum MD, Black KD, Zagotta WN (1995) Molecular mechanism for ligand discrimination of cyclic nucleotide-gated channels. Neuron 15:619–625

Warmke JW, Ganetzky B (1994) A family of potassium channel genes related to *eag* in *Drosophila* and mammals. Proc Natl Acad Sci USA 91:3438–3442

Weber IT, Steitz TA (1987) Structure of a complex of catabolite gene activator protein and cyclic AMP refined at 2.5 Å. J Mol Biol 198:311–326

Wickman K, Nemec J, Gendler SJ, Clapham DE (1998) Abnormal heart rate regulation in GIRK4 knockout mice. Neuron 20:103–114

Wollmuth LP (1995) Multiple ion binding sites in I_h channels of rod photoreceptors in tiger salamander. Pflügers Arch 430:34–43

Wollmuth LP, Hille B (1992) Ionic selectivity of I_h channels of rod photoreceptors in tiger salamanders. J Gen Physiol 100:749–765

Yanagihara K, Irisawa H (1980) Inward current activated during hyperpolarization in the rabbit sinoatrial node cell. Pfluegers Arch 388: 11–19

Yu H, Chang F, Cohen IS (1995) Pacemaker current i_f in adult canine cardiac ventricular myocytes. J Physiol 485:469–483
Zagotta WN, Siegelbaum SA (1996) Structure and function of cyclic nucleotide-gated channels. Annu Rev Neurosci 19:235–263

Editor-in-charge: Professor G. Schultz

Regulation of β-Adrenergic Receptor Responsiveness Modulation of Receptor Gene Expression

S. Danner and M. J. Lohse

Institute of Pharmacology, University of Würzburg, Versbacher Straße 9,
D-97078 Würzburg, Germany

Contents

Abbreviations

βAR, β-adrenergic receptor; βARB, β-adrenergic receptor mRNA binding protein; ARE, AU-rich element; AUF1, AU-rich element RNA-binding/degradation factor 1; bp, base pair(s); CRE, cAMP response element; G_S, stimulatory G-protein; Gβγ, G-protein βγ-subunits; GRE, glucocorticoid response element; GRK, G-protein-coupled receptor kinase; nt., nuctleotide(s); PKA, protein kinase A; PKC, protein kinase C, TRE, thyroid hormone response element; TSS, transcriptional start site; UTR, untranslated region

1 Introduction

β-adrenergic receptors (βAR) are prototypical members of the family of G-protein-coupled receptors, which comprise a large group of seven-transmembrane-helix cell surface receptors for such diverse stimuli as light (Khorana, 1992; Hargrave and McDowell, 1992), hormones and neurotransmitters (Dohlman et al. 1991; Lohse, 1993) and olfactory stimuli (Lancet, 1986; Buck and Axel, 1991). In particular, βAR mediate metabolic and neurocrine actions of the endogenous catecholamines, adrenaline and noradrenaline, as well as of a wide variety of synthetic ligands. Because of their widespread tissue distribution – the β_1AR is the predominant subtype in the heart, the β_2AR is primarily expressed in liver, lung, and smooth muscles, and the β_3AR is highly abundant in brown adipose tissue – and their ability to couple to well-defined effector-systems they serve as a model for the investigation of transmembrane signalling. Upon agonist stimulation βAR couple to the stimulatory G-protein, G_S, which in turn activates the adenylyl cyclase leading to an increase in intracellular cAMP concentrations and subsequently to an activation of protein kinase A (PKA; Hausdorff et al. 1990; Collins et al. 1991; Dohlman et al. 1991; Lohse, 1993). All three components of this signal transduction cascade are subject to complex regulation on both mRNA and protein levels. To date, three βAR-subtypes, termed β_1AR, β_2AR, and β_3AR, have been cloned and sequenced from different species.

In order to discriminate stimuli over a wide concentration range of the respective ligand many G-protein-coupled receptors share the ability to be desensitized, an adaptation process by which they become refractory to further stimulation after an initial response, despite the continuous presence of a stimulus of constant intensity (Hausdorff et al. 1990; Lohse, 1993). Phenomenologically, desensitization can be classified along serveral lines:

mostly important (1) according to the causative stimulus (generalized/heterologous *versus* receptor-specific/homologous), (2) according to the time frame (rapid *versus* slow), or (3) depending on the type of regulatory mechanism involved (impaired receptor function/uncoupling *versus* loss of receptor number/down-regulation). Over the past years, remarkable advances in the understanding of densitization and regulation of βAR have been achieved, however, mainly focusing on regulation of receptor function (Hausdorff et al. 1990; Collins et al. 1991; Dohlman et al. 1991; Lohse, 1993). Therefore, this review emphasizes how changes in βAR gene expression that take place at different levels of regulation – transcription, mRNA stability and translation – create a dynamic and versatile environment for the regulation of βAR receptor responsiveness.

II
Desensitization of Receptor Function

Much of the knowledge about receptor desensitization is based on studies of the β2AR-subtype, which may be due to the fact that the β2AR was the first G-protein-coupled receptor which was purified in significant quantities (Benovic et al. 1984), and the first (apart from rhodopsin) whose primary sequence was elucidated (Dixon et al. 1986; Kobilka et al. 1987a). This receptor shows a pronounced desensitization behaviour (Fig. 1). The individual mechanisms involved in the regulation of β2AR responsiveness are listed in Table 1, together with their respective half-lives and the extent to which they can reduce receptor signalling. Some of these pathways are also responsible for desensitization of other G-protein-coupled receptors, but on the other hand there are many exceptions which are not subject to all of these mechanisms. Therefore, it appears that one reason for the existence of multiple receptor subtypes may be their distinct regulatory properties, since otherwise their responses upon agonist stimulation would be basically the same.

A
βAR Phosphorylation – Functional Uncoupling

Agonist-induced loss of receptor function occurs within seconds after receptor stimulation (Table 1) and is mediated primarily by uncoupling of the receptors from their G-proteins. Such uncoupling can be brought about by the action of two types of kinases: (1) specific kinases, the G-protein-coupled receptor kinases (GRK), and (2) the effector kinases of the βAR system, the protein kinases A and C (PKA, PKC).

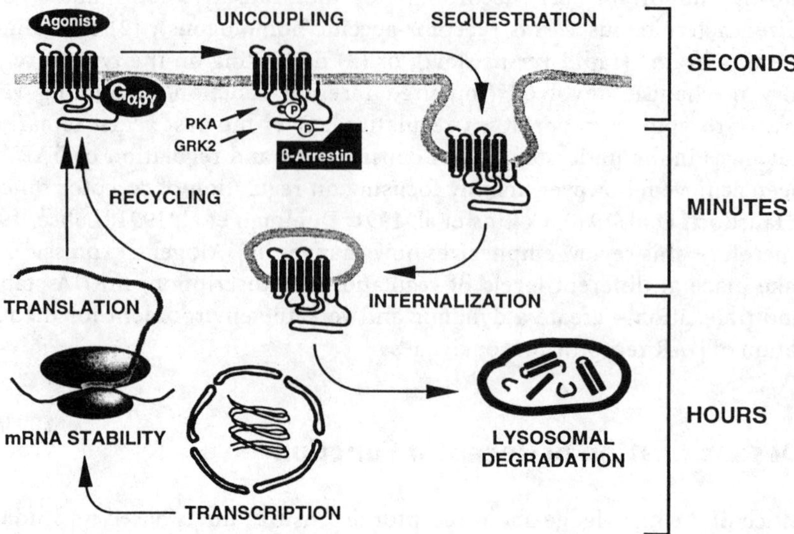

Fig. 1. Schematic illustration of the mechanisms involved in β_2AR desensitization. Agonist stimulation of β_2AR triggers binding and activation of G_s and subsequently of the effector molecule, adenylyl cyclase. To attenuate the signal, the receptors can be functionally uncoupled from their G-proteins by subsequent phosphorylation, binding of β-arrestin and sequestration via clathrin-coated pits (short-term response). After internalization in endosomal vesicles most of the β_2AR are recycled to the plasma membrane. Prolonged or repeated agonist stimulation induces a decrease in β_2AR number (receptor down-regulation; long-term response), which can either be due to enhanced lysosomal degradation of the receptor protein or reduced receptor synthesis caused by changes in transcription rate, mRNA stability or translational regulation, respectively

A.1
Desensitization via Receptor-Specific Kinases

GRKs only phosphorylate agonist-occupied, i.e. active receptors (Benovic et al. 1986). To date, six members of this family have been identified and characterized in molecular detail: GRK1 (rhodopsin kinase), GRK2/3 (β-adrenergic receptor kinases 1/2) and GRK4/5/6 (Lefkowitz, 1993; Lohse et al. 1996; Palczewski, 1997). Even though all six kinases appear to be capable to phosphorylate the β_2AR *in vitro*, only two, GRK2 and 3, have been demonstrated to be involved in the *in vivo* regulation of receptor function. Three serine (Ser-396, Ser-401, Ser-407) and one threonine residues (Thr-384) at the β_2AR C-terminus have been identified as GRK2-phosphorylation sites *in*

Table 1. Mechanisms of β_2AR desensitization. Listed are the individual mechanisms that have been proposed to be responsible for receptor desensitization, together with their onset of occurrence ($t_{1/2}$), and the extent to which they reduce receptor-mediated signalling

Mechanism	**Half life t½**	**Extent (% of receptor function)**
Receptor uncoupling:		
GRK2/β-arrestin	0.1–1 min	50–70
PKA	1–3 min	20–50
Receptor down-regulation:		
protein degradation	0.5–48 h	30–50
red. transcription	0.5–24 h	40–50
mRNA-destabilisation	0.5–24 h	40–50

vitro (Fredericks et al. 1996), but their importance *in vivo* still remain controversial (Seibold et al. 1998).

An absolute requirement for GRK activity is their association with the plasma membrane (Stoffel et al. 1997). In the case of GRK2 and 3, this membrane-association is triggered by activation of the β_2AR. The two kinases bind Gβγ-subunits and are thereby targeted to their membrane-bound receptor substrates (Haga and Haga, 1992; Pitcher et al. 1992a). The Gβγ-binding site of GRK2 has been mapped to a 125-amino acid stretch near its C-terminus (Koch et al. 1993). More recent data indicate that certain phospholipids, in particular phosphatidylinositol 4,5-bisphosphate, are also required for membrane targeting of GRK2 (Touhara et al. 1995; Pitcher et al. 1996).

The capacity of βAR to activate G_s is not markedly altered by GRK-mediated phosphorylation, but phosphorylation, on the other hand, increases the affinity of the β_2AR for β-arrestin (Lohse et al. 1990b; Palczewski, 1994; Sterne-Marr and Benovic, 1995). Binding of β-arrestin to the phosphorylated receptors inhibits β_2AR/G_s-coupling, thereby creating the desensitized state of the receptor (Lohse et al. 1992). The validity of this model is supported by the observation that overexpression of either GRK2 or β-arrestin enhances homologous desensitization of β_2AR (Pippig et al. 1993). Furthermore, Gurevich et al. (1997) demonstrated that β_2AR/β-arrestin complexes have a higher affinity for agonists than receptors alone. Two isoforms of β-arrestin with several splice variants have been identified (Lohse et al. 1990b; Attramandal et al. 1992). However, several studies suggested that β-arrestin2, like GRK3, may predominantly be responsible for desensitization of olfactory receptors (Dawson et al. 1993; Schleicher et al. 1993; Boekhoff et al. 1994). Both PKA and PKC appear to be involved in this regulation, possibly acting in a sequential fashion with GRK3.

With respect to rapid receptor desensitization the β_1AR appears to be remarkably similar to the β_2AR (Freedman et al. 1995). Although certain intracellular domains of the β_1AR differentiate it from the β_2AR at the level of G_s-coupling and sequestration, the primary structures relevant to GRK-mediated phosphorylation are indeed similar. In contrast, the β_3AR exhibited a much less pronounced short-term desensitization or is even resistant to it (Liggett et al. 1993; Nantel et al. 1993; Jockers et al. 1996). The latter study using β_3/β_2AR-chimeras revealed that determinants in the carboxy-terminal tail, the second and the third intracellular loops of the β_2AR provided additive contributions to receptor desensitization, whose interplay has not been investigated in detail so far.

A.2
Desensitization via Effector Kinases

βAR phosphorylation by the effector kinases PKA and PKC provides a negative feedback system, in which PKA controls its own activation. In addition, this mechanism allows a generalized (i.e. heterologous) form of desensitization, since phosphorylation is independent of agonist occupancy of the substrate receptors. Thus, activation of the kinases by any pathway is sufficient to cause phosphorylation (Benovic et al. 1985). Two PKA-consensus sites have been identified in the β_2AR (Blake et al. 1987; Bouvier et al. 1987; Clark et al. 1989), one in the third intracellular loop in a region essentiell for G_s-coupling, and another in the N-terminal part of the C-terminus (Strader et al. 1987; O'Dowd et al. 1988; Münch et al. 1991). However, the site in the third intracellular loop is the clearly preferred one. Phosphorylation in this region seems to be *per se* sufficient to interfere with G-protein activation (Okamoto et al. 1991; Pitcher et al. 1992b). PKC appears to phosphorylate the β_2AR at the same sites that are utilized by PKA (Bouvier et al. 1987; Pitcher et al. 1992). β-arrestins are apparently not involved in heterologous desensitization (Lohse et al. 1992; Pitcher et al. 1992b).

Under optimal conditions, the extent of heterologous desensitization is comparable to that mediated by the homologous pathway (Table 1). Nevertheless, there are two important differences: PKA-mediated desensitization is considerably slower than the GRK-catalyzed pathway ($t_{1/2}$ 2 min *versus* 15 s; Roth et al. 1991), but is much more sensitive to agonist concentrations (half-maximal effects at 10 nM *versus* 300 nM; Hausdorff et al. 1989; Lohse et al. 1990a). Therefore, it is tempting to speculate that PKA-mediated desensitization of the β_2AR is a sensitive ubiquitous process, whereas the receptor-specific pathway might occur at synapses, where high agonist concentrations are present (Lohse et al. 1990a). Indeed, both GRKs as well as β-arrestins are concentrated at postsynaptic compartments (Attramandal et al. 1992; Arriza et al. 1992).

The number of PKA-phosphorylation sites within the different βAR-subtypes – two for the β_2AR, one for the β_1AR, none for the β_3AR – corresponds to the extent of regulation: the β_2-subtype shows a pronounced heterologous desensitization, the β_1AR a much reduced effect, and the β_3-subtype virtually no response (Nantel et al. 1993; Freedman et al. 1995). Another level of complexity in the regulation of βAR-signalling was added by the observation that GRK2 is subject to phosphorylation by PKC (Chuang et al. 1995; Winstel et al. 1996), which connects the desensitization feedback loop mediated by GRK2 with the effector pathway of PKC.

B
Receptor Sequestration

βAR can also become physically uncoupled from G_s by agonist-induced removal from the cell surface and translocation to intracellular compartments, a process called sequestration (Fig. 1). The best evidence for actual internalization of sequestered receptors was obtained by immunofluorescence confocal microscopy (von Zastrow and Kobilka, 1992; von Zastrow et al. 1993), which showed βAR immunoreactivity associated with intracellular vesicles (endosomes).

While originally sequestration was thought to be a mechanism of receptor desensitization, more recent data suggest that its main function might be to resensitize receptors (Sibley et al. 1986; Lohse et al. 1990a; Roth et al. 1991). Receptors appear to be dephosphorylated in intracellular vesicles, and recycled back to the cell surface. Two studies confirmed this hypothesis (Yu et al. 1993; Pippig et al. 1995). The authors showed that blockade of sequestration prevented both receptor dephosphorylation and resensitization. It was concluded that β_2AR sequestration and subsequent recycling to the cellular surface serves to restore the function of desensitized receptors.

The steps following sequestration which lead to resensitization are essentially unknown. A specific membrane-associated G-protein-coupled receptor phosphatase, a member of the PP-2A phosphatase family, has previously been identified (Pitcher et al. 1995). Furthermore, it was shown that β_2AR dephosphorylation requires a specific receptor conformation induced by vesicular acidification (Krueger et al. 1997). Sequestration might facilitate the translocation of β2AR to an acidified compartment, where a conformationally altered receptor associates with the phosphatase, becomes dephosphorylated and subsequently recycled to the plasma membrane (Hausdorff et al. 1989; Lohse et al. 1990a; Yu et al. 1993; von Zastrow and Kobilka, 1994; Pippig et al. 1995).

The molecular determinants of the β2AR involved in sequestration are still unknown. Sequestration is initiated by receptor activation, but early data indicated that it does not seem to require receptor phosphorylation either by GRK2 or by PKA (Strader et al. 1987; Hausdorff et al. 1989; Lohse et al. 1990a). More recent studies, however, suggest that GRK-mediated phosphorylation plus binding of β-arrestin might be a mechanism of receptor sequestration. This hypothesis is based on the observations (1) that overexpression of either GRK2 or β-arrestin rescued the reduced sequestration behaviour of sequestration-defective receptor mutants (Ferguson et al. 1995, 1996), and (2) that β-arrestin can act as a clathrin adaptor, and that this indirect binding of clathrin moves the receptors into clathrin-coated pits (Goodman et al. 1996, 1997). Phosphorylation of Ser-412 in the C-terminus of β-arrestin may regulate its endocytotic function (Lin et al. 1997), since cytoplasmatic β-arrestin is constitutively phosphorylated and translocated to the plasma membrane upon β2AR activation. At the membrane, β-arrestin is rapidly dephosphorylated and this permits binding to clathrin. Additionally, Zhang et al. (1996, 1997) provided evidence that expression levels of β-arrestin are cell-specific.

An alternative sequestration pathway was recently suggested by a β2AR mutant lacking a di-leucine motif in its C-terminus which showed a marked reduction in sequestration (Gabilondo et al. 1997). Di-leucines have been shown to play a role in the intracellular trafficking of many proteins (Letourneur and Klausner, 1992) and appear to represent binding sites for the AP1/AP2 clathrin adaptor proteins (Heilker et al. 1996). The relative importance of the β-arrestin- *versus* the AP1/AP2-dependent pathway may well be cell-specific and influenced by a given cellular milieu.

Whereas the β2AR shows a marked sequestration, this process is dramatically reduced or even abolished in the case of the two other βAR-subtypes (Suzuki et al. 1992; Nantel et al. 1993; von Zastrow et al. 1993; Rousseau et al. 1996). Consequently, the subtype-specific internalization and trafficking properties may also contribute to the specificity of βAR-mediated signalling (Neer and Clapham, 1988).

III
Regulation of Receptor Number

Another mechanism leading to βAR desensitization is a decrease in receptor number, a process also termed receptor down-regulation. While receptor function is regulated rapidly (over seconds to minutes), regulation of receptor number takes much longer, commonly many hours. In isolated cells it is often maximal only after 24 h of continuous agonist exposure. Changes in βAR number can be effected by two classes of mechanisms (Fig. 1): enhanced proteolytic receptor degradation and decreased receptor synthesis, i.e. modulation of gene expression (section IV). Whereas regulation of receptor *function* always results in decreased activity, receptor *expression* can either be up- or down-regulated. Since the mechanisms affecting βAR number operate superimposed on the basal receptor turnover, they are often difficult to quantify separately.

A
βAR Turnover

Receptor expression is the result of a dynamic steady state determined by the rates of degradation and synthesis (Mahan et al. 1987). Under basal conditions most studies in isolated cells have indicated a βAR turnover with half-lives of receptor degradation and synthesis of 24 h, but *in vivo* values of several days have been observed. It appears that receptor recovery following agonist- or cAMP-dependent down-regulation occurs at a faster rate than basal receptor turnover (Mahan et al. 1985; Hughes and Insel, 1986), although the opposite has also been reported (Neve and Molinoff, 1986). In growing cultures receptor turnover is the same or slightly larger than the growth rate of the cells. βAR half-lives between 30 h and 200 h have been determined (Mahan et al. 1987). In some cell lines βAR half-lives increase with the degree of confluence of the cultures (Mahan et al. 1987).

In whole animals receptor turnover is comparably slow. For example, basal turnover of β_1AR and β_2AR in rat heart and lung was found to have half-lives between 5 days and three weeks. However, turnover was somewhat faster in young compared to old animals (Mahan et al. 1987). In rat renal cortex recovery of β_2AR from down-regulation has a $t_{1/2}$ of 18 h, that of the β_1AR of 45 h (Snavely et al. 1985). Therefore, basal turnover of βAR seems to be too slow to play a major role in rapid regulation of receptor expression. As a further consequence, decreases in receptor synthesis probably take even longer – at least *in vivo* – than 24 h (Table 1) to affect receptor numbers. Thus, under most circumstances alterations in receptor mRNA-levels will only affect receptor levels over very extended periods of time.

B
Receptor Down-Regulation

The processes underlying proteolytic receptor degradation are still poorly understood. In some cases, βAR recovery has been shown to require *de novo* protein synthesis, while in other cases it has not (Benovic et al. 1988). However, two components appear to be involved in β_2AR degradation: one which is induced only by agonists, i.e. appears to require agonist occupancy of the receptors, and another mediated by PKA (Bouvier et al. 1989).

B.1
Agonist-Dependent Pathway

The agonist-dependent pathway was documented in mouse S49 lymphoma cells that either lack PKA activity (*kin*$^-$) or display perturbed G_S-effector coupling (*H21a*), but are nevertheless capable of β_2AR down-regulation (Shear et al. 1976; Su et al. 1980; Mahan et al. 1985; Allen et al. 1989). Mutants with impaired receptor-G_S coupling (*cyc*$^-$, *unc*), however, show a blunted, but not a completely blocked down-regulation (Shear et al. 1976; Su et al. 1980; Mahan et al. 1985; Hadcock et al. 1989a; Campbell et al. 1991). These findings raised the possibility that β_2AR-G_S interactions were more important for triggering receptor down-regulation than activation of adenylyl cyclase. This issue was addressed by Campbell et al. (1991) using a series of β_2AR mutants displaying different degrees of impairment in G_S-coupling. Their ability to undergo agonist-mediated down-regulation indeed appeared to reflect their capacities to physically couple to G_S. In the presence of cAMP-inhibitors, no increase in the extent of down-regulation was observed, which implies that the primary defect was not lack of PKA activation. While there is evidence that a G_S-mediated, but PKA-independent pathway is required for β_2AR degradation, the underlying mechanisms, possibly preferential degradation of receptor-G_S complexes, are unknown.

B.2
PKA-Dependent Pathway

The participation of second messengers in β_2AR down-regulation was originally demonstrated in hamster DDT_1-MF2 smooth muscle cells, in which prolonged exposure to βAR-agonists or cAMP-analogs resulted in a loss of receptor binding sites, accompanied by a substantial decrease in β2AR mRNA levels (Hadcock and Malbon, 1988a; Collins et al. 1989). cAMP triggers β2AR down-regulation probably via two independent pathways: cAMP-dependent degradation of the receptor protein itself, and cAMP-dependent reduction of the receptor mRNA which will be discussed in later sections.

In order to examine the role of cAMP in promoting β2AR down-regulation Bouvier et al. (1989) compared wild-type β2AR with mutants in which one or both consensus sites for PKA phosphorylation were substituted. Upon activation, cAMP-induced phosphorylation of the mutant β2ARs was completely abolished, but down-regulation was only moderately decreased compared to the wild-type receptor. Thus, the delay in the rate of down-regulation induced by cAMP is most likely due to alterations of receptor phosphorylation, although phosphorylation is certainly not the only factor contributing to this regulatory process. As discussed above, early studies suggested that GRK-dependent phosphorylation did not influence receptor downregulation (Strader et al. 1987; Campbell et al. 1991). However, since the GRK/β-arrestin pathway seems to be involved in receptor sequestration (see above) and since at least some studies suggest that sequestration is a prerequisite for subsequent receptor degradation (von Zastrow and Kobilka, 1992), it seems plausible that GRKs also play a role in receptor downregulation.

The other two βAR-subtypes exhibit markedly different patterns of down-regulation: β1AR show only a modest, β3AR little or no down-regulation. In Chinese hamster fibroblasts transfected with β1AR cDNAs no receptor down-regulation was detected in the first 4 h of agonist treatment, compared with an about 50% reduction of β2AR (Suzuki et al. 1992). After 24 h 60% of β1AR were still present, whereas less than 20% of β2AR remained. Similar results were obtained in human neuroepithelioma cells (Fishman et al. 1991) and in 3T3-F442A adipocytes, respectively (Thomas et al. 1992). In the case of the β3AR, down-regulation is not only blunted but receptor expression has been found to be actually increased to about 160% of basal expression (Thomas et al. 1992; Nantel et al. 1993), which indicates a unique regulatory mode for this subtype, whose physiological relevance still remains unclear.

IV
Regulation of βAR Gene Expression

Because of their function in signal transduction it appears plausible that the gene expression of βAR is also tightly regulated in response to changes in agonist concentrations. This hypothesis was originally confirmed in hamster DDT$_1$-MF2 smooth muscle cells as well as in rat C6 glioma cells, in which the agonist-induced reduction of βAR numbers was closely associated with a decrease of the corresponding mRNA levels (Hadcock and Malbon, 1988a; Hadcock et al. 1989b; Collins et al. 1989, 1990).

Such down-regulation of mRNA levels can occur even without receptor occupancy by agonists. In contrast to receptor agonists, cAMP treatment led to a more modest decrease in β$_2$AR mRNA. A correlation between activation of the PKA-dependent pathway and a reduction of β$_2$AR mRNA concentration was demonstrated by Hadcock et al. (1989a) for various mouse S49 lymphoma mutant cell lines. Agonist stimulation reduced β$_2$AR mRNA levels in wild-type cells by about 50%, whereas down-regulation was unaffected in *kin*$^-$ cells. In the β$_2$AR-G$_S$ coupling mutants, *cyc*$^-$ and *unc*, receptor mRNA levels were decreased only upon activation of PKA by forskolin. The result in mutant *H21a* was, however, surprising, since a significant reduction of β$_2$AR mRNA levels was determined even though in these cells there is no coupling of receptor activation to an increase in cAMP levels.

Bouvier et al. (1989) also reported a considerable decrease of the corresponding β$_2$AR mRNA levels of both the wild-type and mutant receptors, which preceded the reduction of βAR numbers. Down-regulation of β$_2$AR mRNA levels in these cells occurred in the absence of the endogenous promoter, suggesting that posttranscriptional mechanisms must play an important role in the regulation of β$_2$AR gene expression.

Today we know that – like regulation of the receptor protein itself – multiple mechanisms are involved in the regulation of the receptor mRNA. These involve both alterations in gene transcription and regulation at the posttranscriptional, i.e. mRNA- level. In order to understand these various mechanisms, the structure and function of the respective genes and mRNAs need to be known.

A
Gene Structure

βAR-subtypes show striking similarities in their gene structures and therefore appear to have evolved from a common ancestor gene. One of the most surprising characteristics of both the β$_1$AR and β$_2$AR genes is their lack of introns, a very unusual feature of eukaryotic genes (Dixon et al. 1986; Kobilka et al. 1987b; Frielle et al. 1987; Emorine et al. 1989; Machida et al. 1990). Most other adrenergic receptors also lack introns in their genes, with the exceptions of the α_{1A}AR and the α_{1B}AR which have been shown to contain introns (Ramarao et al. 1992; Perez et al. 1994). The existence of an intron was also observed within the gene of the turkey β_1AR, which differs from its mammalian counterpart in pharmacological properties (Wang and Ross, 1995). Removal of this intron resulted in a 59 amino acid-residue extension of the receptor which blocked its ability to undergo agonist-mediated endocytosis (Wang and Ross, 1995).

In contrast to the β_1- and β_2AR genes, the β_3AR genes of man and rodents contain two protein-coding exons (Nahmias et al. 1991; Granneman et al. 1992, 1993; van Spronsen et al. 1993; Granneman and Lahners, 1994). Furthermore, the mouse and rat genes contain an additional intron in the 3′-untranslated region (3′UTR) of the gene. The introns interrupt the coding region twelve (rat, mouse) or six amino acids (man) from the carboxy-terminus of the receptor. They may contain enhancer elements that could be important for the predominant adipose tissue specific expression of this gene. In addition, a sequence region within the intron of the human β_3AR gene is highly homologous to the second exon of the mouse β_3AR gene. Whether this observation has any physiological significance, however, remains unclear. Likewise, the possible function of the second intron in the 3′UTR of the rodent β_3AR genes is unknown (Granneman et al. 1992, 1993; van Spronsen et al. 1993).

Multiple transcriptional start sites (TSS) have been identified in all three βAR-subtypes (Fig. 2): in the human and rat β_1AR genes they are located between positions −270 and −220 relative to the start codon (Searles et al. 1995; Evanko et al. 1998), in the mouse β_1AR gene between positions −450 and −350 (Cohen et al. 1993); transcriptional start sites of the human, hamster, and rat β_2AR genes are found approximately 250 bp upstream of the start codon, in the case of the rat gene an additional TSS is located at position −60 (Kobilka et al. 1987b; Emorine et al. 1987; McGraw et al. 1996; Jiang et al. 1996; Baeyens et al. 1998); and the start sites of rodent and human β_3AR genes are located between positions −200 and −150 (van Spronsen et al. 1993; Granneman and Lahners, 1994).

Alternative polyadenylation sites have been reported in the 3′UTRs of all these sequences which explain the different transcript sizes obseverd. β_1AR transcripts vary between 2.5 kb and 3.0 kb in length (Frielle et al. 1987; Feve et al. 1990; Machida et al. 1990; Cohen et al. 1993; Evanko et al. 1998), for the β_2AR two mRNA species have been detected – a major one ranging between 2.0 kb and 2.2 kb in size and a minor one between 1.6 kb and 1.8 kb (Emorine et al. 1987; Collins et al. 1988, 1989; Baeyens and Cornett, 1993; Jiang et al. 1996). β_3AR transcripts between 2.1 kb and 2.8 kb have been found in several human and rodent tissues (Emorine et al. 1989; van Spronsen et al. 1993; Krief et al. 1993; Granneman and Lahners, 1994). However, in

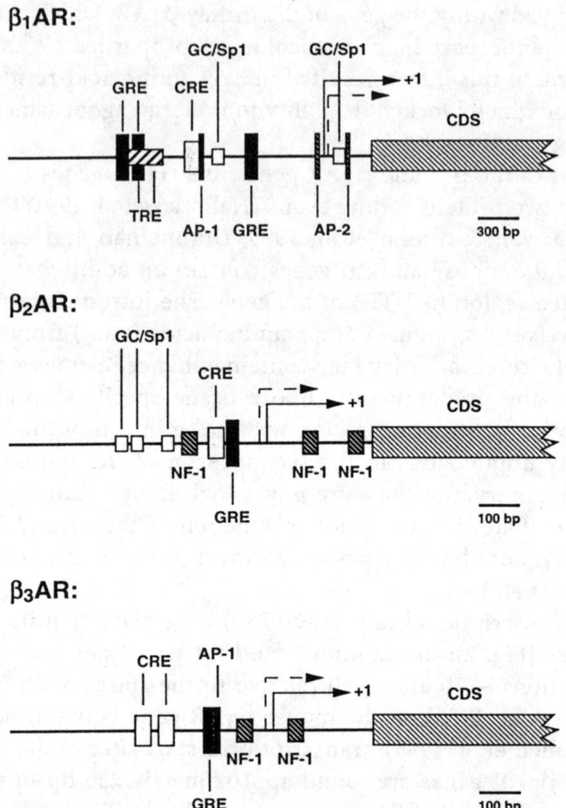

Fig. 2. Comparison of the proximal promoter regions of the three human βAR subtypes. The βAR coding regions are indicated by hatched boxes, the transcriptional start sites as arrows. The regulatory elements depicted are referred to in the text (see also Table 2)

some human organs larger mRNA species were also visible (Emorine et al. 1989).

B
Promoter Organization

The βAR promoter regions share many general characteristics of promoters for so-called housekeeping genes, i.e. genes that are constitutively expressed at low levels. These include a high G+C-content, the lack of canonical TATA and CAAT boxes and a high frequency of the CpG dinucleotides (Bird, 1986;

Table 2. Comparison of the gene structures/promoter regions of the three βAR subtypes based on the respective sequences of man, rat, mouse, and hamster available so far. TF, transcription factor; n.d., not determined. Numbers in brackets indicate the numbers of the respective sites

Sequence feature	β_1AR	β_2AR	β_3AR
Gene structure:			
Introns(1)	(–)	–	+(1–2)
5'ORF(2)	(–)	+	–
Promoter region:			
CRE	+(1)	+(1)	+(1–4)
GRE	+(2–3)	+(≥ 5)	+(1)
TRE	+(2)	+(1)	n.d.
TATA-box	–	–	–
CAAT-box(3)	(–)	(–)	(–)
GC-box	+	+	+
TF binding sites:			
CREB/CREM	+(1)	+(1)	+(1–4)
AP-1	+(1)	n.d	+(1)
AP-2	+(3)	+(4)	n.d
Sp1	+(3)	+(3–4)	n.d
others	AP-4	CP-1, C/EBPα	NF–1
	ICER	NF-1, NF-κB	

(1) The turkey β1AR subtype consists of two exons and an alternatively spliced intron (Wang and Ross, 1995). (2) Two small ORFs of unknown function have been reported within the rat, mouse, and human β1AR 5′UTRs (Machida et al. 1990; Cohen et al. 1993; Evanko et al. 1998). (3) All three βAR subtypes posses CAAT-box approximations but no canonical sequence motifs.

Smale and Baltimore, 1989; Zawel and Reinberg, 1993). The proximal promoter regions of the three human βAR genes are schematically shown in Fig. 2, and the most important features that are present in the human, rat, mouse, and hamster sequences are summarized in Table 2.

Deletion analyses of the β1AR and β2AR 5′ regions demonstrated promoter activity for at least two separate domains: the primary promoter of the human β1AR gene is located between positions –440 and –360 (Evanko et al. 1998). Two repressor elements and one transcriptional enhancer have been identified between positions −3000 and −2000 and around position −1900, (Evanko et al. 1998). In the rat sequence the basal promoter activity is located around 50 nt. downstream compared to the human promoter (Searles et al. 1995; Bahouth et al. 1997b). Again, two repressor elements have been identified (positions −2800 and −120), the latter interestingly 3′ of the TSS. These discrepancies suggested that expression and transcriptional regulation of the β_1AR gene are regulated in a species-specific manner.

The basal promoters of the human and rat β_2AR genes are both located between positions −400 and −100 (Kobilka et al. 1987b; Emorine et al. 1987; McGraw et al. 1996; Jiang et al. 1996; Baeyens et al. 1998). Jiang et al. (1996) reported two alternative promoters between −300/−180 and −100/−40 and an additional repressor element around position −1000. No deletion analyses are currently available for the β_2AR gene. However, based on the location of the TSS at ≈ −250 and ≈ −60 (see above) the primary promoter should be localized also around positions −300 to −100.

Gene expression of all three βAR-subtypes is tightly regulated by steroid hormones (glucocorticoids, thyroid hormones) as well as by cAMP. Different 'response elements' (GRE, TRE, and CRE; Evans, 1988; Roesler et al. 1988) recognized by the respective hormone receptors and cAMP-dependent transcription factors, respectively, are found in various numbers and locations within the βAR promoter sequences (Fig. 2 and Table 2). They are discussed in more detail in the following section.

Another characterisitc feature of βAR promoters is their high G+C-content. In the β_1AR and β_2AR genes several binding sites for the transcription factor Sp1(GGGCGG; Dynan and Tjian, 1985) have been identified (Kobilka et al. 1987b; Emorine et al. 1987; Collins et al. 1993; Cohen et al. 1993; Searles et al. 1994; Jiang and Kunos, 1995; Jiang et al. 1996). Additionally, all these promoters are bound by members of the CREB/CREM-family of transcription factors (Habener et al. 1995) and by AP-1/AP-2 (activating protein-1/-2), dimeric transcription factors composed of Jun, Fos or ATF (Karin et al. 1997). Footprinting analyses and sequence compilation studies revealed that the βAR promoter regions are targets of a series of different transcription factors, including AP-4, CP1, NF-κB, NF-1, and C/EBPα (Collins et al. 1993; van Spronsen et al. 1993; Searles et al. 1994; Jiang and Kunos, 1995; Jiang et al. 1996). The selective binding of these different *trans*-acting factors may be involved in the tissue-specific and developmental regulation of βAR gene expression.

C
Regulation of βAR Transcription

C.1
Steroid Hormones

Some of the first studies to address transcriptional control of βAR focused on regulation by glucocorticoids and thyroid hormones. Glucocorticoids increase β_2AR numbers and adenylyl cyclase activities 2–5 fold depending on the cell-type investigated (Hadcock and Malbon, 1988b; Malbon and Hadcock, 1988; Collins et al. 1988; Nakada et al. 1989; Feve et al. 1990), an effect preceded by a rapid elevation of the corresponding mRNA. A direct increase in the β_2AR gene transcription rate (approximately threefold) was shown to be the mechanistic basis of this observation (Collins et al. 1988), since the half-life of the β_2AR transcript remained unchanged even after prolonged hormonal treatment (Hadcock and Malbon, 1988b; Hadcock et al. 1989b). Sequence elements within the 5´UTR of the β_2AR gene (GREs are also present in the coding region and the 3´UTR) with great homology to the 15 bp glucocorticoid response element (GRE) consensus sequence (Evans, 1988) were identified as the responsible *cis*-acting elements (Malbon and Hadcock, 1988; Nakada et al. 1989; Cohen et al. 1993; McGraw et al. 1996). Treatment of DDT_1-MF2 smooth muscle cells with a combination of both β_2AR-agonists and dexamethasone, a synthetic glucocorticoid, demonstrated a dynamic regulation and adaptation of receptor gene expression to different hormonal signals. The agonist-mediated down-regulation of both β_2AR numbers and mRNA levels could be reversed by glucocorticoid treatment (Hadcock et al. 1989b). The enhanced rate of gene transcription was sufficient to overcome down-regulation, and *vice versa* β_2AR up-regulation after steroid treatment was significantly lowered by addition of a receptor agonist. This finding is also of clinical importance since therapy of asthmatics with βAR-agonists to relax bronchial smooth muscles over time results in receptor refractoriness. In a recent study (Mak et al. 1995), it was shown that glucocorticoids can prevent β_2AR down-regulation at the transcriptional level without affecting β_1AR. Interestingly, the transcription factor CREB appeared to be involved in this phenomenon.

In contrast to the β_2AR, the β_1-subtype is down-regulated by about 50% upon glucocorticoid treatment due to a suppression of β_1AR gene transcription (Guest et al. 1990; Feve et al. 1990; Kiely et al. 1994). However, the detailed mechanistic basis for this effect has not been elucidated so far. In 3T3-F442A adipocytes long-term exposure to dexamethasone not only led to a significant reduction of β_1AR gene expression but also to a sharp 4–8 fold decrease of both β3AR mRNA steady-state levels and β3AR numbers (Feve et al. 1992). The glucocorticoid receptor was identified as the major mediator of this inhibition, presumably by negative interaction with the activator protein-1 (AP-1) and/or additional tissue-specific factors (Feve et al. 1990, 1992).

βAR gene expression can also be modulated by thyroid hormones which may be the basis for the many sympathomimetic effects accompanying hyperthyroidism (Nakada et al. 1989; Lazar-Wesely et al. 1991). However, thyroid hormones modulate the adrenergic effects of catecholamines in a βAR-subtype- and tissue-specific manner, an observation that is not understood in terms of its mechanisms. In cultured rat cardiac myocytes, β1AR mRNA levels are up-regulated about threefold by thyroid hormones, whereas the β2AR mRNA levels remain unaffected (Bahouth, 1991). The thyroid hormone-promoted regulation of βAR gene expression occurs via activation of a thyroid hormone response element (TRE; Evans, 1988) located in the 5′-flanking region of the receptor gene. Bahouth et al. (1997a) further demonstrated that the β1AR TRE is unusual in that it is a direct repeat separated by five instead of four nucleotides and that it is located 3′ to the TSS. This underscores again the specificity of βAR transcriptional regulation.

C.2
cAMP

Gene expression of the different βAR-subtypes is also regulated by cAMP, in a process referred to as 'autoregulation', i.e. the ability of the receptor/effector complex to directly regulate the transcription of its own genes, which has been most thoroughly investigated for the β2AR gene so far (Hough and Chuang, 1990; Collins et al. 1989, 1990; Hosoda et al. 1994, 1995). Short-term exposure (≤ 30 min) to β-agonists or cAMP analogs resulted in a 3- to 5-fold elevation of β2AR mRNA levels due to a direct transcriptional activation of the β2AR gene, without any effects on transcript stability at these early time points (Collins et al. 1989). The increase in mRNA concentration is transient and followed, upon prolonged β2AR stimulation, by a typical pattern of down-regulation, accompanied by a reduction of βAR numbers and changes in mRNA turnover which will be discussed later (Collins et al. 1989, 1990). These dynamic alterations in both the transcription rate and mRNA stability, represent a complex regulatory paradigm.

The transcriptional response of the β2AR gene to cAMP is mediated by binding of the 'cAMP response element binding protein' (CREB), a 43 kDa-phosphoprotein (Habener et al. 1995) to the respective response element

(CRE; Roesler et al. 1988), GTACGTCA, in the β_2AR promoter region (Kobilka et al. 1987b; Collins et al. 1990; McGraw et al. 1996). In a recent report, Rohlff et al. (1997) provided evidence that Sp1, one of the predominant transcription factors modulating β_2AR gene expression, is regulated by PKA. Therefore, it is tempting to speculate that stimulation of Sp1 may also contribute to the observed cAMP-induced enhancement of the β_2AR transcription rate. However, the physiological significance of this cAMP-promoted up-regulation still remains unclear.

This biphasic regulatory pattern described above, with short-term agonist treatment increasing and prolonged treatment decreasing the levels of receptor mRNA, holds also true for the β_1-subtype (Hough and Chuang, 1990; Hosoda et al. 1994). The latter authors demonstrated that the β_1AR mRNA down-regulation depends on *de novo* protein synthesis suggesting the participation of an inducible, short-lived protein component. In the case of the β_1AR the half-life of the receptor transcript was unaffected, which in turn led to the conclusion that the reduction in mRNA levels is indeed mediated by a cAMP-induced decrease in β_1AR transcription. Consistent with these results was the identification of CREs in the human, rat, and mouse β_1AR promoter regions (Collins et al. 1993; Cohen et al. 1993; Searles et al. 1994, 1995). It is tempting to speculate that stimulation of the β_1AR results in a PKA-dependent phosphorylation of members of the CREB/CREM-family of transcription factors, which in turn activates receptor gene expression. Rydelek-Fitzgerald et al. (1996) have identified an 'inducible cAMP early repressor' (ICER) in rat C6 glioma cells, whose gene expression is rapidly induced upon agonist-stimulation of the β_1AR. It appears that ICER binds to the CRE in the β_1AR promoter region and thereby causes transcriptional repression.

Contradicting results have been reported for the cAMP-dependent regulation of β_3AR gene expression. Thomas et al. (1992) detected a dramatic increase (about 65%) in β_3AR mRNA levels upon chronic agonist exposure in the white adipocyte cell-line 3T3-F442A, probably caused by the presence of four CREs within the β_3AR 5′-flanking region. On the other hand, in brown fat cells a transient (24 h) 50% down-regulation of β_3AR transcript concentrations was observed, which was further shown to be due to a cAMP-mediated cessation of transcription and not caused by an increased mRNA-instability (Bengtsson et al. 1996). These findings represent another example of a cell type-specific regulation.

C.3
Cell-Type Specificity

Transcriptional control of gene expression appears to be the only pathway regulating the β_1AR- and the β_3AR-subtypes. In the case of the β_2AR the situation is more complex, and regulation of this mRNA appears to occur in a cell type-specific manner: in cells with a vast majority of β_1AR, such as C6 glioma cells, β_2AR mRNA down-regulation occurs via a reduction of the transcription rate (Hough and Chuang, 1990; Collins et al. 1989, 1990; Hosoda et al. 1995; Danner and Lohse, 1997), whereas in cells with a high portion of β_2AR, e.g. DDT$_1$-MF2 smooth muscle cells, mRNA levels are modulated at the posttranscriptional level via a decrease in transcript stability (Hadcock and Malbon, 1988a; Hadcock et al. 1989b; Danner and Lohse, 1997). However, the extent of down-regulation is largely comparable in both cell lines. Upon agonist-stimulation β_2AR mRNA levels were reduced by 50%, but the receptor numbers were decreased by about 80%. This demonstrates that, due to degradation of the receptor protein, receptor down-regulation exceeds the extent of mRNA reduction (Bouvier et al. 1989; Danner and Lohse, 1997). Elevation of cAMP levels with forskolin also caused a 50% reduction in β_2AR mRNA levels, which led to the conclusion that in both cell types β_2AR mRNA regulation is essentially mediated by cAMP. In contrast to isoproterenol, however, forskolin reduced receptor numbers by only 50%, i.e. by the same extent as it reduced receptor mRNA levels. These data suggest that the larger receptor down-regulation seen with agonists ($\approx$ 80%) compared to forskolin ($\approx$ 50%) is cAMP-independent and probably due to degradation of the receptor protein.

Determination of the receptor mRNA half-lives in DDT$_1$-MF2 cells showed a 50% reduction after agonist-treatment, i.e. in this cell-type β_2AR mRNA down-regulation can be solely explained by transcript destabilization (Hadcock and Malbon, 1988a; Hadcock et al. 1989b; Danner and Lohse, 1997). In contrast, the β_2AR half-life in C6 glioma cells was completely unaffected by agonists (Hosoda et al. 1995; Danner and Lohse, 1997). In both cases – i.e. inhibition of mRNA transcription and destabilization of the mRNA – inducible short-lived protein components appear to be involved: in DDT$_1$-MF2 cells, the extent of β_2AR mRNA destabilization was significantly reduced (about 50%) by addition of exotoxin A to prevent *de novo* protein synthesis, whereas in C6 glioma cells the agonist-induced down-regulation of the receptor transcript was completely abolished by exotoxin A treatment (Hosoda et al. 1995; Danner and Lohse, 1997). These results clearly underline a cell type-specific regulation of β_2AR gene expression at either transcriptional or posttranscriptional levels. The molecular factors that determine the choice of the respective regulatory pathway in an individual cell have not been identified to date. However, in rat liver both regulatory mechanisms participate in modulating β2AR gene expression during development (Baeyens and Cornett, 1993, 1995; Baeyens et al. 1998). The hepatic β2AR density declines progressively during postnatal development. Both reduction of the transcription rate mediated by a repressor protein and a decrease in mRNA-stability equally contribute to this regulatory pattern.

D
Regulation of β_2AR mRNA-Stability

First hints on the regulation of β2AR mRNA down-regulation at the post-transcriptional level were obtained by Bouvier et al. (1989) who showed that agonist-mediated down-regulation occurs even if the receptor gene is under the control of a cAMP-independent promoter. Malbon and coworkers (Hadcock et al. 1989) provided direct evidence for β2AR transcript destabilization in DDT1-MF2 cells in which the receptor mRNA half-life decreased by about 50% upon stimulation with isoproterenol. Since the half-life was determined after transcriptional blockade by actinomycin D, this reduction must be due to an increased mRNA turnover. Since then it turned out that posttranscriptional control of gene expression is widely distributed among G-protein-coupled recepors, such as the α1BAR (Izzo et al. 1990), the m_1-muscarinic acetylcholine receptor (Lee et al. 1994) and the angiotensin AT1 receptor (Lassegue et al. 1995; Wang et al. 1997, Thekkumkara et al. 1998). Again, the prototypical receptor in such studies is the β2AR, whereas for the two other βAR-subtypes no regulation via changes in mRNA-stability has been observed so far.

Posttranscriptional mechanisms are of particular interest, since they participate in the stability and turnover of various highly labile mRNAs, such as the transcripts of granulocyte/macrophage colony stimulating factor (GM-CSF), cytokines and the proto-oncogenes c-*fos* and c-*myc* (Sachs, 1993; Ross, 1995; Chen and Shyu, 1995; Jacobson and Peltz, 1996). AU-rich elements (AREs) are often found within the 3′UTRs of these mRNAs and appear to be key determinants of their short half-lives (Caput et al. 1986). The sequence features of these motifs are summarized in Table 3. Their functional importance was demonstrated by Shaw and Kamen (1986) who observed a heterologous destabilization of the normally highly stable β-globin mRNA after insertion of an ARE from GM-CSF into the β-globin 3′UTR. Although AREs are actually the predominant class of mRNA stability determinants, mRNA turnover does not strictly depend on these motifs (Ross, 1995; Jacobson and Peltz, 1996). For example, Brown et al. (1996) defined a new class of stem-

Table 3. Structural and functional features of the three classes of AREs (modified according to Chen and Shyu, 1995). The sequence elements represent only minimal consensus motifs and can vary considerably in an individual case. A biphasic mRNA decay kinetics means that deadenylation precedes the degradation of the mRNA body. Thereby, synchronous deadenylation results in decay intermediates with poly(A) tails of 30–60 residues, whereas asynchronous (progressive) deadenylation leads to the formation of completely deadenylated mRNAs

ARE	Sequence feature, mRNA decay kinetics	Example
AUUUA-containing Class I	1–3 scattered copies of AUUUA-motifs additional U-rich regions biphasic; synchronous deadenylation	c-*fos* c-*myc*
AUUUA-containing Class II	at least two overlapping copies of the nonamer UUAUUUA(U/A)(U/A) embedded in a U-rich region biphasic; asynchronous deadenylation	GM-CSF IL-3
Non-AUUUA	U-rich sequences additional still unknown features biphasic; asynchronous deadenylation	c-*jun*

loop containing destabilization motifs present in a variety of cytokine-mRNAs. Furthermore, in the c-*fos* and c-*myc* mRNAs determinants within the coding sequence also contribute to transcript stability (Kabnick and Housman, 1988; Shyu et al. 1991; Herrick and Ross, 1994). In the case of c-*fos* these determinants have been shown to function independently from each other and from the AREs within the 3′UTR (Wellington et al. 1993; Schiavi et al. 1994).

The exact sequence requirements of an ARE are a matter of intensive research. Based on two recent *in vitro* studies with artificial sequences, the minimal ARE destabilization motif was suggested to be UUAUUUA(U/A)(U/A) (Lagnado et al. 1994; Zubiaga et al. 1995). Two copies of this nonamer efficiently destabilize β-globin reporter mRNAs. However, there is also evidence that an AUUUA pentamer need not be an integral part of a functional ARE (Chen and Shyu, 1994; Peng et al. 1996). On the other hand, very recent studies indicate that a reiteration of AUUUA-pentamers is the essential ARE sequence motif (Xu et al. 1997). Therefore, it appears that each ARE represents a combination of structurally and functionally distinct domains, such as AUUUA motifs, AU-nonamers and U-rich elements, and that it is the combination of these sequence elements, which determines its ultimate destabilizing potency (Chen et al. 1994; Chen and Shyu, 1995).

AREs appear to represent the recognition sites for several cytoplasmic and nuclear-associated RNA-binding proteins, generally 30–50 kDa in size, which either mediate mRNA degradation (Brewer, 1991; Vakalopoulou et al. 1991; Bohjanen et al. 1991, 1992; Zhang et al. 1993; Myer et al. 1997) or prevent mRNA decay by 'masking' the ARE (Malter, 1991; Rajagopalan and Malter, 1994). Although some of these ARE-binding proteins have been purified, their precise roles in the regulation of mRNA-stability and turnover have not been elucidated so far.

Several AREs have also been identified in the 3´UTRs of different G-protein-coupled receptor mRNAs, in particular within the β_2AR transcripts of man, rat, and hamster, respectively (Dixon et al. 1986; Kobilka et al. 1987a; Huang et al. 1993; Baeyens and Cornett, 1995; Tholanikunnel et al. 1995; Pende et al. 1996; Tholanikunnel and Malbon, 1997; Danner et al. 1998). Interestingly, none of them exactly fits the proposed ARE consensus motif mentioned above (Lagnado et al. 1994; Zubiaga et al. 1995). Additionally, pronounced species-specific differences in their sequence composition as well as localization make a general regulatory pattern very unlikely (Fig. 3). However, AREs appear to play an important role also in destabilizing the β_2AR mRNA upon agonist-stimulation.

Three binding proteins have been described for the β_2AR mRNA so far (Fig. 3): (1) the 'β-adrenergic receptor mRNA binding protein'(βARB), a M_r 35,000 cytosolic protein presumably specific for the β_2AR which was identified in hamster DDT$_1$-MF2 smooth muscle cells whose concentration varies inversely with the level of receptor mRNA (Port et al. 1992; Huang et al. 1993); (2) a M_r 85,000 factor regulating β_2AR transcript destabilization in adult rat hepatocytes which, however, appears to be restricted to these cells because of its unusual size and its association with β_2AR-mediated glucose metabolism in liver (Baeyens and Cornett, 1995; Baeyens et al. 1998); and (3) the M_r 37,000/40,000 'AU-rich element RNA-binding/degradation factor' (AUF1), which has also been shown to bind the β_1AR 3´UTR (Pende et al. 1996). AUF1 is the only one of these factors which has been cloned (Zhang et al. 1993; Ehrenman et al. 1994; Wagner et al. 1998), and one of the few RNA-binding proteins for which a causal function in mRNA decay has been demonstrated (Brewer, 1991; Zhang et al. 1993). AUF1 binds an ARE presumably as a hexameric protein via its two RNA recognition motifs (RRM; Burd and Dreyfuss, 1994) and additional N- and C-terminal sequence domains (DeMaria et al. 1997). Furthermore, it was shown that its ARE binding affinity correlates very well with the potency of an ARE to direct mRNA degradation (DeMaria and Brewer, 1996).

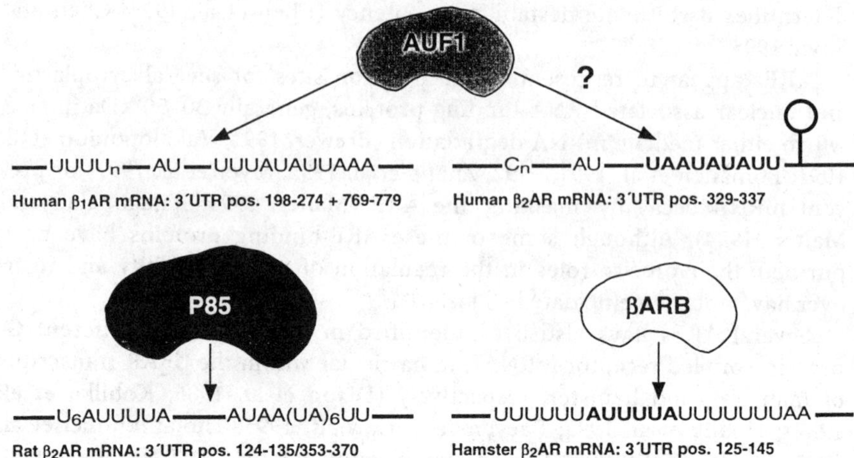

Fig. 3. RNA-binding proteins involved in the regulation of β2AR mRNA stability. Potential protein binding motifs within the respective β2AR 3´UTRs are indicated. For the two AREs shown in bold protein binding has been demonstrated *in vitro* (Tholanikunnel and Malbon, 1997; Danner et al. 1998). The nucleotide following the stop codon was numbered with 1 in all four 3´UTR sequences. βARB, β-adrenergic receptor binding protein; AUF1, AU-rich element RNA-binding/degradation factor 1; P85, 85 kDa βAR mRNA-binding protein

Binding of all three β2AR mRNA-specific proteins is selectively reduced by poly(U) RNA (Port et al. 1992; Baeyens and Cornett, 1995; Pende et al. 1996). Furthermore, *in vitro* binding of βARB to the hamster β2AR mRNA requires both an AUUUA pentamer and U-rich flanking domains (Huang et al. 1993). Since neither the hamster nor the human β2AR 3´UTRs possess any AU-rich consensus motifs, it is tempting to speculate that agonist-induced β2AR mRNA destabilization occurs via unique *cis*-acting elements. This hypothesis was confirmed in two recent studies (Tholanikunnel and Malbon, 1997; Danner et al. 1998), in which a 20 nucleotide AU-rich domain with an unusual AUUUUA hexamer core in the 3´UTR of the hamster transcript, and a non-canonical AU-nonamer in the human β2AR mRNA 3´UTR, respectively, were shown to be obligate for agonist-induced β2AR mRNA destabilization (Fig. 3). Tholanikunnel and Malbon (1997) demonstrated that the 20 nt. ARE at positions 125–145 of the 3´UTR functions as the binding site for βARB *in vivo*. Interestingly, substitution of the rare hexamer by an AUUUA-pentamer resulted in a 50% reduction of βARB binding activity, which paralles results of Bohjanen et al. (1992) with the AU-B protein and underscores the specificity of this regulatory element. U to G substitutions that interrupt the hexamer and/or the 5´- or 3´-flanking U-rich domains of this ARE reduced or even abolished agonist-induced destabilization of the hamster β_2AR mRNA.

In the human β_2AR mRNA, the nonamer UAAUAUAUU at positions 329–337 of the 3´UTR represents the destabilization element. This nonamer differs from the proposed ARE consensus sequence at positions 2 and 5 (A for U in both cases) which both have been shown to be important for the general destabilizing potency of AREs (Lagnado et al. 1994; Zubiaga et al. 1995). However, mutation of these positions to the consensus sequence did not result in enhanced destabilization of the receptor mRNA (Danner et al. 1998). Since the region, in which the ARE is embedded, does not resemble the U-rich sequences found in other highly labile mRNAs (Ross, 1995; Chen and Shyu, 1995), the deviations from the ARE consensus may be compensated by other β_2AR-specific elements. These might be located in two additional AU-rich domains within the 3´UTR and/or secondary structure elements, such as a predicted stem-loop immediately downstream of the AU-nonamer.

The importance of this ARE in the human β_2AR mRNA was comfirmed by the observation that a protein (or a protein complex) selectively bound to the 3´ half of the human receptor transcript (Danner et al. 1998). The β_2AR mRNA-protein interaction had a sequence specificity identical to that found in destabilization experiments, and protein binding can be selectively suppressed by an excess of oligomer comprising the AU-nonamer (Danner and Lohse, unpublished). The synthesis of this mRNA binding factor(s) was induced in DDT$_1$-MF2 smooth muscle cells by β_2AR-stimulation. However, its nature has not been elucidated so far. It is tempting to speculate that it may be identical to AUF1 which has been shown to bind both human β_1AR and β_2AR 3´UTRs and whose expression is induced in DDT$_1$-MF2 cells upon β_2AR stimulation (Pende et al. 1996). The biochemical and functional relatedness of AUF1 and βARB led initially to the assumption that they might be identical, but immunochemical experiments suggested that they were distinct (Pende et al. 1996)

The analysis of a β-globin/β_2AR 3´UTR chimeric transcript demonstrated that elements encoded in the human β_2AR 3´UTR are not only necessary but also sufficient for mRNA destabilization. Cloning of the β_2AR 3´UTR behind the coding sequence of the normally stable β-globin gene resulted in a cAMP-sensitive chimeric mRNA (Danner et al. 1998). Although the stability of the chimeric mRNA was only about one third of the wild-type β-globin transcript, increases in cAMP with either isoproterenol or forskolin caused a further 2-fold decrease of the mRNA half-life. The almost identical results obtained with isoproterenol and forskolin suggest a predominant role for cAMP as a regulator of β_2AR mRNA stability. However, the biochemical mechanisms mediating this cAMP-dependent regulation of β2AR mRNA stability remain to be elucidated.

Gene expression of the two other βAR-subtypes, β1AR and β3AR, appears to be exclusively regulated on the transcriptional level (see section IV.C). To date, there are no reports available postulating any participation of posttranscriptional mechanisms. However, Pende et al. (1996) demonstrated that AUF1 also interacts with the 3´UTR of the β1AR mRNA. This observation is intriguing since AUF1 and its corresponding mRNA are up-regulated in the failing human heart, in which both β1AR numbers and β1AR mRNA levels are reduced by about 50% (Ungerer et al. 1993; Bristow et al. 1993). The questions whether an increase in AUF1 protein is associated with an increased β1AR mRNA turnover in the failing heart or whether AUF1 indeed regulates β1AR mRNA-stability still remain to be answered.

E
Regulation of β_2AR mRNA Translation

In addition to transcriptional and posttranscriptional control mechanisms, β2AR gene expression can also be regulated on the translational level (Fig. 4). In the 5´UTR of the receptor mRNA a small open reading frame (sORF) was identified encoding a 19 amino acid peptide (β2AR upstream peptide, BUP), which is highly conserved in the human, rat, mouse, and hamster sequences, (Dixon et al. 1986; Kobilka et al. 1987a). The ATG triplet at position −110 serving as the putative start codon of the sORF is, in contrast to

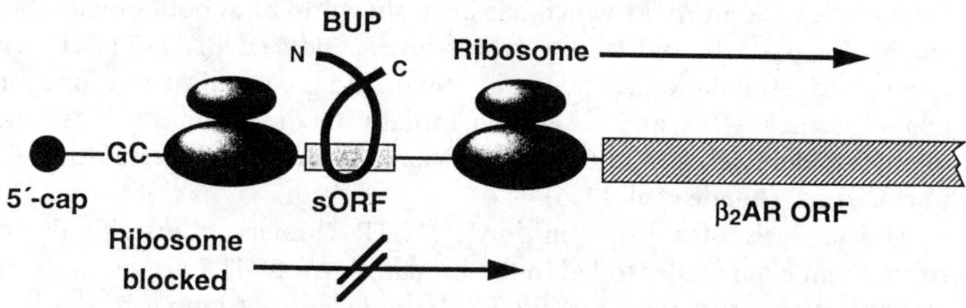

Fig. 4. Regulation of β2AR mRNA translation. Model of the inhibition of β2AR mRNA translation by the β2AR upstream peptide (BUP), which is encoded by a small open reading frame (sORF) within the β2AR 5´UTR (adapted from Parola and Kobilka, 1994). The GC-rich element upstream of the sORF enhances mRNA recognition by the ribosome. After translation the nascent BUP binds ist own mRNA and thereby prevents ribosome scanning past this complex. The β2AR ORF is translated by ribosomes that either scan past the sORF start codon without initiation or that reinitiate at the receptor start codon after translating the sORF

the β2AR start codon, in a poor sequence context for translational initiation (Kozak, 1984, 1991). However, as revealed by a –110ATG → CCT substitution translation initiates at the sORF ATG as efficiently as at the receptor start codon (Parola and Kobilka, 1994). It has been speculated that a GC-rich sequence element located 5′ to the sORF (positions –173 to –117 of the human β2AR mRNA), which is highly conserved in mammalian β2AR transcripts, may function as a translational enhancer. β2AR mRNAs lacking this upstream open reading frame were found to be translated almost ten times more efficiently than the wildtype transcript both in an *in vitro* rabbit reticulocyte system and in *Xenopus* oocytes (Kobilka et al. 1987c). Mutational inactivation of the sORF start codon increased β2AR expression about twofold in transiently transfected COS-7 cells (Parola and Kobilka, 1994). Furthermore, it could be demonstrated that mutations within the 5′UTR increasing translation of the sORF comcomitantly decrease β2AR translation. Additional mutations in the peptide coding region suggested that BUP translation is a requirement for inhibition of receptor expression (Parola and Kobilka, 1994). The same authors proposed a model in which this peptide binds its own mRNA and thereby interferes in ribosome-mRNA interactions, possibly by preventing ribosomes from scanning past the BUP-mRNA complex. The β2AR open reading frame is translated by ribosomes that either scan past the sORF start codon without initiation (leaky scanning) or by ribosomes that reinitiate after translation of the sORF. Alternatively, the nascent peptide may directly bind to the ribosme, which subsequently impedes translational initiation at the downstream receptor cistron. The specificity of this inhibition may be achieved by the high local concentration of BUP (threefold stoichiometric excess compared to β2AR). On the other hand, BUP is not a 'classic' *trans*-acting factor at low peptide concentrations. Dissociation of the peptide from the transcript would relieve translational inhibition. For the two other βAR subtypes no comparable regulation at translational level has been reported so far.

The 50% decrease in β2AR translation observed in COS-7 cells caused by this translational mechanism is similar in magnitude to the agonist-induced reduction of the β2AR transcription rate in C6 glioma cells (Collins et al. 1989, 1990; Hosoda et al. 1995; Danner and Lohse, 1997), as well as the changes in β2AR mRNA-stability upon prolonged receptor stimulation in DDT1-MF2 smooth muscle cells (Hadcock et al. 1989b; Tholanikunnel and Malbon, 1997; Danner and Lohse, 1997; Danner et al. 1998). Since β2AR gene expression in general appears to be regulated in a cell type-specific manner, one might assume that translational modulation is the essential regulatory pathway in COS-7 cells. However, it is unclear, whether BUP is always translated with the same efficiency compared to the β2AR. Future studies will have to elucidate how the distinct but probably additive pathways interact and whether they can be observed in a single cell. Furthermore, the factors which determine how β2AR responsiveness is primarily regulated under specific conditions remain to be identified.

V
Conclusions

Chronic stimulation of βAR results in a decrease of receptor responsiveness, a process called agonist-induced receptor desensitization. Distinct mechanisms operating on both mRNA and protein levels contribute to this complex regulatory network. The three βAR-subtypes differ markedly in their general desensitization properties – β2 receptors display a pronounced, β1 receptors only a modest and β3 receptors little or no regulation, which may be one of the reasons for the existence of different receptor subtypes. Furthermore, not all of the regulatory cascades described are utilized in a given setting. It has previously turned out that cell-type specific mechanisms are used depending on the capacity of an individual cell or tissue to synthesize the protein components necessary to elicit or to modulate βAR desensitization as well as on the quality of the desensitizing stimulus. Recent findings concerning both β1AR transcriptional regulation as well as β2AR mRNA stability add an additional level of complexity to this system since the transcripts of various species differ significantly in their *cis*-acting sequence elements, suggesting distinct regulatory pathways.

Undoubtly, the elucidation of the molecular mechanisms responsible for the regulation of βAR responsiveness is only beginning to be unravelled. There are still many questions to be answered which concern not only the mode of activation, but also of inactivation of these receptors. Little is known about the proteins that mediate intracellular sorting and trafficking of βAR, and the intriguing variability of modulation of β2AR gene expression is just emerging.

So far, βAR signalling and regulation have been extensively studied *in vitro*. However, the *in vivo* relevance of the individual pathways remains unclear, so that generation and analysis of transgenic mice will further help to elucidate the regulation and the physiological or pathophysiological role played by the different βAR-subtypes and the various pathways that can regulate their function and expression.

Acknowledgements. We thank Dr. Lutz Hein for critical reading of the manuscript. Work in the authors´ laboratory is supported by grants from the Deutsche Forschungsgemeinschaft (SFB 355), the Bundesministerium für Bildung und Forschung, and the Fonds der Chemischen Industrie.

References

Allen, J.M., Abrass, I.B., and Palmiter, R.D. (1989) β_2-adrenergic receptor regulation after transfection into a cell line deficient in the cAMP-dependent protein kinase. *Mol. Pharmacol.* 36, 248–255.

Arriza, J.L., Dawson, T.M., Simerly, R.B., Martin, L.J., Caron, M.G., Snyder, S.H., and Lefkowitz, R.J. (1992) The G-protein-coupled receptor kinases βARK1 and βARK2 are widely distributed at synapses in rat brain. *J. Neurosci.* 12, 4045–4055.

Attramadal, H., Arriza, J.L., Aoki, C., Dawson, T.M., Codina, J., Kwatra, M.M., Snyder, S.H., Caron, M.G., and Lefkowitz, R.J. (1992) β-arrestin2, a novel member of the arrestin/β-arrestin gene family. *J. Biol. Chem.* 267, 17882–17890.

Baeyens, D.A., and Cornett, L.E. (1993) Transcriptional and posttranscriptional regulation of hepatic β_2-adrenergic receptor gene expression during development. *J. Cell. Physiol.* 157, 70–76.

Baeyens, D.A., and Cornett, L.E. (1995) Association of hepatic β_2-adrenergic receptor gene transcript destabilization during postnatal development in the Sprangue-Dawley rat with a M_r 85,000 protein that binds selectively to the β_2-adrenergic receptor mRNA 3′-untranslated region. *J. Cell. Physiol.* 163, 305–311.

Baeyens, D.A., McGraw, D.W., Jacobi, S.E., and Cornett, L.E. (1998) Transcription of the β_2-adrenergic receptor gene in rat liver is regulated during early postnatal development by an upstream repressor element. *J. Cell. Physiol.* 175, 333–340.

Bahouth, S.W. (1991) Thyroid hormones transcriptionally regulate the β_1-adrenergic receptor gene in cultured ventricular myocytes. *J. Biol. Chem.* 266, 15863–15869.

Bahouth, S.W., Cui, X., Beauchamp, M.J., and Park, E.A. (1997a) Thyroid hormone induces β_1-adrenergic receptor gene transcription through a direct repeat separated by five nucleotides. *J. Mol. Cell. Cardiol.* 29, 3223–3237.

Bahouth, S.W., Cui, X., Beauchamp, M.J., Shimomura, H., George, S.T., and Park, E.A. (1997b) Promoter analysis of the rat β_1-adrenergic receptor gene identifies sequences involved in basal expression. *Mol. Pharmacol.* 51, 620–629.

Bengtsson, T., Redegren, K., Strosberg, A.D., Nedergaard, J., and Cannon, B. (1996) Down-regulation of β_3 adrenoceptor gene expression in brown fat cells is transient and recovery is dependent upon a short-lived protein factor. *J. Biol. Chem.* 271, 33366–33375.

Benovic, J.L., Bouvier, M., Caron, M.G., and Lefkowitz, R.J. (1988) Regulation of adenylyl cyclase-coupled β-adrenergic receptors. *Annu. Rev. Cell. Biol.* 4, 405–428.

Benovic, J.L., Pike, L.J., Cerione, R.A., Staniszewski, C., Yoshimasa, T., Codina, J., Caron, M.G., and Lefkowitz, R.J. (1985) Phosphorylation of the mammalian β-adrenergic receptor by cyclic AMP-dependent protein kinase. *J. Biol. Chem.* 260, 7094–7101.

Benovic, J.L., Caron, M.G., and Lefkowitz, R.J. (1984) The mammalian β_2-adrenergic receptor: Purification and characterization. *Biochemistry* 23, 4519–4525.

Benovic, J.L., Strasser, R.H., Caron, M.G., and Lefkowitz, R.J. (1986) β-adrenergic receptor kinase: Identification of a novel protein kinase that phosphorylates the agonist-occupied form of the receptor. *Proc. Natl. Acad. Sci. USA* 83, 2797–2801.

Bird, A.P. (1986) CpG-rich islands and the function of DNA methylation. *Nature* 321, 209–213.

Blake, A.D., Mumford, R.A., Strout, H.V., Slater, E.G., and Strader, C.G. (1987) Synthetic segments of the mammalian β-adrenergic receptor are preferentially recognized by cAMP-dependent protein kinase and protein kinase C. *Biochem. Biophys. Res. Commun.* 147, 168–173.

Boekhoff, I., Inglese, J., Schleicher, S., Koch, W.J., Lefkowitz, R.J., and Breer, H. (1994) Olfactory desensitization requires membrane targeting of receptor kinase mediated by βγ-subunits of heterotrimeric G proteins. *J. Biol. Chem.* 269, 37–40.

Bohjanen, P.R., Petryniak, B., June, C.H., Thompson, C.B., and Lindsten, T. (1991) An inducible cytoplasmatic factor (AU-B) binds selectively to AUUUA multimers in the 3′ untranslated region of lymphokine mRNA. *Mol. Cell. Biol.* 11, 3288–3295.

Bohjanen, P.R., Petryniak, B., June, C.H., Thompson, C.B., and Lindsten, T. (1992) AU RNA-binding factors differ in their binding specifities and affinities. *J. Biol. Chem.* 267, 6302–6309.

Bouvier, M., Collins, S., O′Dowd, B.F., Campbell, P.T., de Blasi, A., Kobilka, B.K., MacGregor, C., Irons, G.P., Caron, M.G., and Lefkowitz, R.J. (1989) Two distinct pathways for cAMP-mediated down-regulation of the β_2-adrenergic receptor. Phosphorylation of the receptor and regulation of its mRNA level. *J. Biol. Chem.* 264, 16786–16792.

Bouvier, M., Leeb-Lundberg, L.M.F., Benovic, J.L., Caron, M.G., and Lefkowitz, R.J. (1987) Regulation of adrenergic receptor function by phosphorylation. II. Effects of agonist occupancy on phosphorylation of α_1- and β_2-adrenergic receptors by protein kinase C and the cyclic AMP-dependent protein kinase. *J. Biol. Chem.* 262, 3106–3113.

Brewer, G. (1991) An A+U-rich element RNA-binding factor regulates c-*myc* mRNA stability in vitro. *Mol. Cell. Biol.* 11, 2460–2466.

Bristow, M.R., Minobe, W.A., Raynolds, M.V., Port, D.J., Rasmussen, R., Ray, P.E., and Feldman, A.M. (1993) Reduced β_1 receptor messenger abundance in the failing human heart. *J. Clin. Invest.* 92, 2737–2745.

Brown, C.Y., Lagnado, C.A., and Goodall, G.J. (1996) A cytokine mRNA-destabilizing element that is structurally and functionally distinct from A+U-rich elements. *Proc. Natl. Acad. Sci. USA* 93, 13721–13725.

Buck, L., and Axel., R. (1991) A novel multigene family may encode odorant receptors: A molecular basis for odorant recognition. *Cell* 65, 175–187.

Burd, C.G., and Dreyfuss, G. (1994) Conserved structures and diversity of functions of RNA-binding proteins. *Science* 265, 615–620.

Campbell, P.T., Hnatowich, M., O′Dowd, B.F., Caron, M.G., Lefkowitz, R.J., and Hausdorff, W.P. (1991) Mutations of the human β_2-adrenergic receptor that impair coupling to G_s interfere with receptor down-regulation but not sequestration. *Mol. Pharmacol.* 39, 192–198.

Caput, D., Beutler, B., Hartog, K., Thayer, R., Brown-Shimer, S., and Cerami, A. (1986) Identification of a common nucleotide sequence in the 3′-untranslated regions of mRNA molecules specifying inflammatory mediators. *Proc. Natl. Acad. Sci. USA* 83, 1670–1674.

Chen, C.Y.A., Chen, T.M., and Shyu, A.B. (1994) Interplay of two functionally and structurally distinct domains of the c-*fos* AU-rich element specifies its mRNA-destabilizing function. *Mol. Cell. Biol.* 14, 416–426.

Chen, C.Y.A., and Shyu, A.B. (1994) Selective degradation of early-response-gene mRNAs: Functional analyses of sequence features of the AU-rich elements. *Mol. Cell. Biol.* 14, 8471–8482.

Chen, C.Y.A., and Shyu, A.B. (1995) AU-rich elements: Characterization and importance in mRNA degradation. *Trends Biochem. Sci.* 20, 465–470.

Chuang, T.T., LeVine, H., and De Blasi, A. (1995) Phosphorylation and activation of β-adrenergic receptor kinase by protein kinase C. *J. Biol. Chem.* 270, 18660–18665.

Clark, R.B., Friedman, J., Dixon, R.A.F., and Strader, C.D. (1989) Identification of a specific site required for rapid heterologous desensitization of the β-adrenergic receptor by cAMP-dependent protein kinase. *Mol. Pharmacol.* 36, 343–348.

Cohen, J.A., Baggott, L.A., Romano, C., Arai, M., Southerling, T.E., Young, L.A., Kozak, C.A., Molinoff, P.B., and Greene, M.I. (1993) Characterization of a mouse β_1-adrenergic receptor genomic clone. *DNA Cell Biol.* 12, 537–547.

Collins, S., Altschmied, J., Herbsman, O., Caron, M.G., Mellon, P.L., and Lefkowitz, R.J. (1990) A cAMP response element in the β_2-adrenergic reseptor gene confers transcriptional autoregulation by cAMP. *J. Biol. Chem.* 265, 19330–19335.

Collins, S., Bouvier, M., Bolanowski, M.A., Caron, M.G., and Lefkowitz, R.J. (1989) cAMP stimulates transcription of the β_2-adrenergic receptor gene in response to short-term agonist exposure. *Proc. Natl. Acad. Sci. USA* 86, 4853–4857.

Collins, S., Caron, M.G., and Lefkowitz, R.J. (1988) β_2-adrenergic receptors in hamster smooth muscle cells are transcriptionally regulated by glucocorticoids. *J. Biol. Chem.* 263, 9067–9070.

Collins, S., Lohse, M.J., O'Dowd, B., Caron, M.G., and Lefkowitz, R.J. (1991) Structure and regulation of G-protein-coupled receptors: The β_2-adrenergic receptor as a model. *Vitam. Horm.* 46, 1–39.

Collins, S., Ostrowski, J., and Lefkowitz, R.J. (1993) Cloning and sequence analysis of the human β_1-adrenergic receptor 5'-flanking region. *Biochim. Biophys. Acta* 1172, 171–174.

Danner, S., and Lohse, M.J. (1997) Cell type-specific regulation of β_2-adrenoceptor mRNA by agonists. *Eur. J. Pharmacol.* 331, 73–78.

Danner, S., Frank, M., and Lohse, M.J. (1998) Agonist-regulation of human β_2-adrenergic receptor mRNA stability occurs via a specific AU-rich element. *J. Biol. Chem.* 273, 3223–3229.

Dawson, T.M., Arriza, J.L., Jaworsky, D.E., Borisy, F.F., Attramadal, H., Lefkowitz, R.J., and Ronnett, G.V. (1993) β-adrenergic receptor kinase-2 and β-arrestin-2 as mediators of odorant-induced desensitization. *Science* 259, 825–829.

DeMaria, C.T., and Brewer, G. (1996) AUF1 binding affinity to A+U-rich elements correlates with rapid mRNA degradation. *J. Biol. Chem.* 271, 12179–12184.

DeMaria, C.T., Sun, Y., Long, L., Wagner, B.J., and Brewer, G. (1997) Structural determinants in AUF1 required for high affinity binding to A + U-rich elements. *J. Biol. Chem.* 272, 27635–27643.

Dixon, R.A.F., Kobilka, B.K., Strader, D.J., Benovic, J.L., Dohlman, H.G., Frielle, T., Bolanowski, M.A., Bennett, C.D., Rands, E., Diehl, R.E., Mumford, M.A., Slater, E.E., Sigal, I.S., Caron, M.G., Lefkowitz, R.J., and Strader, C.D. (1986) Cloning of the gene and cDNA for mammalian β-adrenergic receptor and homology with rhodopsin. *Nature* 321, 75–79.

Dohlman, H.G., Thorner, J., Caron, M.G., and Lefkowitz, R.J. (1991) Model systems for the study of seven-transmembrane-segment receptors. *Annu. Rev. Biochem.* 60, 653–688.

Dynan, W.S., and Tjian, R. (1985) Control of eukaryotic messenger RNA synthesis by sequence-specific DNA-binding proteins. *Nature* 316, 774–778.

Ehrenman, K., Long, L., Wagner, B.J., and Brewer, G. (1994) Characterization of cDNAs encoding the murine A+U-rich RNA-binding protein AUF1. *Gene* 149, 315–319.

Emorine, L.J., Marullo, S., Briend-Sutren, M.M., Patey, G., Tate, K., Delavier-Klutchko, C., and Strosberg, A.D. (1989) Molecular characterization of the human β_3-adrenergic receptor. *Science* 245, 1118–1121.

Emorine, L.J., Marullo, S., Delavier-Klutchko, C., Kaveri, S.V., Durieu-Trautmann, O., and Strosberg, A.D. (1987) Structure of the gene for human β_2-adrenergic receptor: Expression and promoter characterization. *Proc. Natl. Acad. Sci. USA* 84, 6995–6999.

Evanko, D.S., Ellis, C.E., Venkatachalam, V., and Frielle, T. (1998) Preliminary analysis of the transcriptional regulation of the human β_1-adrenergic receptor gene. *Biochem. Biophys. Res. Commun.* 244, 395–402.

Evans, R.M. (1988) The steroid and thyroid hormone receptor superfamily. *Science* 240, 889–895.

Ferguson, S.S.G., Downey, W.E., Colapietro, A.M., Barak, L.S., Menard, L., and Caron, M.G. (1996) Role of β-arrestin in mediating agonist-promoted G protein-coupled receptor internalization. *Science* 271, 363–366.

Ferguson, S.S.G, Menard, L., Barak, L.S., Koch, W.J., Colapietro, A.M., and Caron, M.G. (1995) Role of phosphorylation in agonist-promoted β_2-adrenergic receptor sequestration. Rescue of a sequestration-defective mutant receptor by βARK1. *J. Biol. Chem.* 270, 24782–24789.

Feve, B., Baude, B., Krief, S., Strosberg, A.D., Pairault, J., and Emorine, L.J. (1992) Inhibition by dexamethasone of β_3-adrenergic receptor responsiveness in 3T3-F442A adipocytes. Evidence for a transcriptional mechanism. *J. Biol. Chem.* 267, 15909–15915.

Feve, B., Emorine, L.J., Briend-Sutren, M.M., Lasnier, F., Strosberg, A.D., and Pairault, J. (1990) Differential regulation of β_1- and β_2-adrenergic receptor protein and mRNA levels by glucocorticoids during 3T3-F442A adipose differentiation. *J. Biol. Chem.* 265, 16343–16349.

Fishman, P.H., Nussbaum, E., and Duman, R.S. (1991) Characterization and regulation of β_1-adrenergic receptors in a human neuroepithelioma cell line. *J. Neurochem.* 56, 596–602.

Fredericks, Z.L., Pitcher, J.A., and Lefkowitz, R.J. (1996) Identification of the G protein-coupled receptor kinase phosphorylation sites in the human β_2-adrenergic receptor. *J. Biol. Chem.* 271, 13796–13803.

Freedman, N.J., Liggett, S.B., Drachman, D.E., Pei, G., Caron, M.G., and Lefkowitz, R.J. (1995) Phosphorylation and desensitization of the human β_1-adrenergic receptor. *J. Biol. Chem.* 270, 17953–17961.

Frielle, T., Collins, S., Daniel, K.W., Caron, M.G., Lefkowitz, R.J., and Kobilka, B.K. (1987) Cloning of the cDNA of the human β_1-adrenergic receptor. *Proc. Natl. Acad. Sci. USA* 84, 7920–7924.

Gabilondo, A.M., Hegler, J., Krasel, C., Boivin-Jahns, V., Hein, L., and Lohse, M.J. (1997) A dileucine motif in the C terminus of the β_2-adrenergic receptor is involved in receptor internalization. *Proc. Natl. Acad. Sci. USA* 94, 12285–12290.

Gabilondo, A.M., Krasel, C., and Lohse, M.J. (1996) Mutations of Tyr^{326} in the β_2-adrenoceptor disrupt multiple receptor functions. *Eur. J. Pharmacol.* 307, 243–250.

Goodman, O.B., Krupnik, J.G., Gurevich, V.V., Benovic, J.B., and Keen, J.H. (1997) Arrestin/clathrin interaction. Localization of the arrestin binding locus to the clathrin terminal domain. *J. Biol. Chem.* 23, 15017–15022.

Goodman, O.B., Krupnik, J.G., Santini, F., Gurevich, V.V., Renn, R.B., Gagnon, A.W., Keen, J.H., and Benovic, J.B. (1996) β-arrestin acts as a clathrin adapter in endocytosis of the β_2-adrenergic receptor. *Nature* 383, 447–450.

Granneman, J.G., and Lahners, K.N. (1994) Analysis of human and rodent β_3-adrenergic receptor messenger ribonucleic acids. *Endocrinology* 130, 109–114.

Granneman, J.G., Lahners, K.N., and Chaudhry, A. (1993) Characterization of the human β_3-adrenergic receptor gene. *Mol. Pharmacol.* 44, 264–270.

Granneman, J.G., Lahners, K.N., and Rao, D.D. (1992) Rodent and human β_3-adrenergic receptor genes contain an intron within the protein-coding block. *Mol. Pharmacol.* 42, 964–970.

Guest, S.J., Hadcock, J.R., Watkins, D.C., and Malbon, C.C. (1990) β_1- and β_2-adrenergic receptor expression in differentiating 3T3-L1 cells. *J. Biol. Chem.* 265, 5370–5375.

Gurevich, V.V., Pals-Rylaarsdam, R., Benovic, J.L., Hosey, M.M., and Onorato, J.J. (1997) Agonist-receptor-arrestin, an alternative ternary complex with high agonist affinity. *J. Biol. Chem.* 272, 28849–28852.

Habener, J.F., Miller, C.P., and Vallejo, M. (1995) cAMP-dependent regulation of gene transcription by cAMP response element-binding protein and cAMP response element modulator. *Vitam. Horm.* 51, 1–57.

Hadcock, J.R., and Malbon, C.C. (1988a) Down-regulation of β-adrenergic receptors: Agonist-induced reduction in receptor mRNA levels. *Proc. Natl. Acad. Sci. USA* 85, 5021–5025.

Hadcock, J.R., and Malbon, C.C. (1988b) Regulation of β-adrenergic receptors by "permissive" hormones: Glucocorticoids increase steady-state levels of receptor mRNA. *Proc. Natl. Acad. Sci. USA* 85, 8415–8419.

Hadcock, J.R., Ros, M., and Malbon, C.C. (1989a) Agonist regulation of β-adrenergic receptor mRNA. Analysis in S49 mouse lymphoma mutants. *J. Biol. Chem.* 245, 13956–13961.

Hadcock, J.R., Wang, H.Y., and Malbon, C.C. (1989b) Agonist-induced destabilization of β-adrenergic receptor mRNA. Attenuation of glucocorticoid-induced up-regulation of β-adrenergic receptors. *J. Biol. Chem.* 264, 19928–19933.

Haga, K., and Haga, T. (1992) Activation by G protein βγ subunits of agonist- or light-dependent phosphorylation of muscarinic acetylcholine receptors and rhodopsin. *J. Biol. Chem.* 267, 2222–2227.

Hargrave, P.A., and McDowell, J.H. (1992) Rhodopsin and phototransduction: A model system for G protein-linked receptors. *FASEB J.* 6, 2323–2331.

Hausdorff, W.P., Bouvier, M., O´Dowd, B.F., Irons, G.P., Caron, M.G., and Lefkowitz, R.J. (1989) Phosphorylation sites on two domains of the β_2-adrenergic receptor are involved in distinct pathways of receptor desensitization. *J. Biol. Chem.* 264, 12657–12665.

Hausdorff, W.P., Caron, M.G., and Lefkowitz, R.J. (1990) Turning off the signal: Desensitization of β-adrenergic receptor function. *FASEB J.* 4, 2881–2889.

Heilker, R., Manning-Krieg, U., Zuber, J.F., and Spiess, M. (1996) *In vitro* binding of clathrin adaptors to sorting signals correlates with endocytosis and basolateral sorting. *EMBO J.* 15, 2893–2899.

Herrick, D.J., and Ross, J. (1994) The half-life of c-*myc* mRNA in growing and serum-stimulated cells: Influence of the coding and 3′ untranslated regions and role of ribosome translocation. *Mol. Cell. Biol.* 14, 2119–2128.

Hosoda, K., Feussner, G.K., Rydelek-Fitzgerald, L., Fishman, P.H., and Duman, R.S. (1994) Agonist and cAMP-mediated regulation of β_1-adrenergic receptor mRNA and gene transcription in rat C6 glioma cells. *J. Neurochem.* 63, 1635–1645 .

Hosoda, K., Fitzgerald, L.R., Vaidya, V.A., Feussner, G.K., Fishman, P.H., and Duman, R.S. (1995) Regulation of β_2-adrenergic receptor mRNA and gene transcription in rat C6 glioma cells: Effects of agonist, forskolin, and protein synthesis inhibition. *Mol. Pharmacol.* 48, 206–211.

Hough, C., and Chuang, D.M. (1990) Differential down-regulation of β_1- and β_2-adrenergic receptor mRNA in C6 glioma cells. *Biochem. Biophys. Res. Commun.* 170, 46–52.

Huang, L.Y., Tholanikunnel, B.G., Vakalopoulou, E., and Malbon, C.C. (1993) The M_r 35,000 β-adrenergic receptor mRNA-binding protein induced by agonists requires both an AUUUA-pentamer and U-rich domains for RNA recognition. *J. Biol. Chem.* 268, 25769–25775.

Hughes, R.J., and Insel, P.A. (1986) Agonist-mediated regulation of $alpha_1$- and $beta_2$-adrenergic receptor metabolism in a muscle cell line, BC3H-1. *Mol. Pharmacol.* 29, 521–530.

Izzo, N.J., Seidman, C.E., Collins, S., and Colucci, W.S. (1990) α_1-adrenergic receptor mRNA level is regulated by norepinephrine in rabbit aortic smooth muscle cells. *Proc. Natl. Acad. Sci. USA* 87, 6268–6271.

Jacobson, A., and Peltz, S.W. (1996) Interrelationships of the pathways of mRNA decay and translation in eukaryotic cells. *Annu. Rev. Biochem.* 65, 693–739.

Jiang, L., and Kunos, G. (1995) Sequence of the 5′ regulatory domain of the gene encoding the rat β_2-adrenergic receptor. *Gene* 163, 331–332.

Jiang, L., Gao, B., and Kunos, G. (1996) DNA elements and protein factors involved in the transcription of the β_2-adrenergic receptor gene in rat liver. The negative regulatory role of C/EBPα. *Biochemistry* 35, 13136–13146.

Jockers, R., Da Silva, A., Strosberg, A.D., Bouvier, M., and Marullo, S. (1996) New molecular and structural determinants involved in β_2-adrenergic receptor desensitization and sequestration. Delineation using chimeric β_3/β_2-adrenergic reeptors. *J. Biol. Chem.* 271, 9355–9362.

Kabnick, K.S., and Housman, D.E. (1988) Determinants that contribute to cytoplasmatic stability of human c-*fos* and β-globin mRNAs are located at several sites in each mRNA. *Mol. Cell. Biol.* 8, 3244–3250.

Karin, M., Liu, Z.G., and Zandi, E. (1997) AP-1 function and regulation. *Curr. Opin. Cell Biol.* 9, 240–246.

Khorana, H.G. (1992) Rhodopsin, photoreceptor of the rod cell. *J. Biol. Chem.* 26, 1–4.

Kiely, J., Hadcock, R.J., Bahouth, S.W., and Malbon, C.C. (1994) Glucocorticoids down-regulate β_1-adrenergic-receptor expression by suppressing transcription of the receptor gene. *Biochem. J.* 302, 397–403.

Kobilka, B.K., Dixon, R.A.F., Frielle, T., Dohlman, H.G., Bolanowski, M.A., Sigal, I.S., Yang-Feng, T.L., Francke, U., Caron, M.G., and Lefkowitz, R.J. (1987a) cDNA for the human β_2-adrenergic receptor: A protein with multiple membrane-spanning domains and encoded by a gene whose chromosomal location is shared with that of the receptor for platelet-derived growth factor. *Proc. Natl. Acad. Sci. USA* 84, 46–50.

Kobilka, B.K., Frielle, T., Dohlman, H.G., Bolanowski, M.A., Dixon, R.A.F., Keller, P., Caron, M.G. and Lefkowitz, R.J. (1987b) Delineation of the intronless nature of the genes for the human and hamster β_2-adrenergic receptor and their putative promoter regions. *J. Biol. Chem.* 262, 7321–7327.

Kobilka, B.K., MacGregor, C., Kiefer, D., Kobilka, T.S., Caron, M.G., and Lefkowitz, R.J. (1987c) Functional activity and regulation of human β_2-adrenergic receptors expressed in *Xenopus* oocytes. *J. Biol. Chem.* 262, 15796–15802.

Koch, W.J., Inglese, J., Stone, W.C., and Lefkowitz, R.J. (1993) The binding site for the βγ subunits of heterotrimeric G proteins on the β-adrenergic receptor kinase. *J. Biol. Chem.* 268, 8256–8260.

Kozak, M. (1984) Compilation and analysis of sequences upstream from the transcriptional start site in eukaryotic mRNAs. *Nucl. Acids Res.* 12, 857–872.

Kozak, M. (1991) Structural features in eukaryotic mRNAs that modulate the initiation of translation. *J. Biol. Chem.* 266, 19867–19870.

Krief, S., Lönnqvist, F., Raimbault, S., Baude, B., Van Spronsen, A., Arner, P., Strosberg, A.D., Ricquier, D., and Emorine, L.J. (1993) Tissue distribution of β_3-adrenergic receptor mRNA in man. *J. Clin. Invest.* 91, 344–349.

Krueger, K.M., Daaka, Y., Pitcher, J.A., and Lefkowitz, R.J. (1997) The role of sequestration in G protein-coupled receptor resensitization. Regulation of β_2-adrenergic receptor dephosphorylation by vesicular acidification. *J. Biol. Chem.* 272, 5–8.

Lagnado, C.A., Brown, C.Y., and Goodall, G.J. (1994) AUUUA is not sufficient to promote poly(A) shortening and degradation of an mRNA: The functional sequence within AU-rich elements may be UUAUUUA(U/A)(U/A). *Mol. Cell. Biol.* 14. 7984–7995.

Lancet, D. (1986) Vertebrate olfactory reception. *Annu. Rev. Neurosci.* 9, 329–355.

Lassegue, B., Alexander, R.W., Nickenig, G., Clark, M., Murphy, T.J., and Griendling, K.K. (1995) Angiotensin II down-regulates the vascular smooth muscle AT_1 receptor by transcriptional and post-transcriptional mechanisms: Evidence for homologous and heterologous regulation. *Mol. Pharmacol.* 48, 601–609.

Lazar-Wesley, E., Hadcock, R.J., Malbon, C.C., Kunos, G., and Ishac, E.J.N. (1991) Tissue-specific regulation of α_{1B}, β_1, and β_2-adrenergic receptor mRNAs by thyroid state in the rat. *Endocrinology* 129, 1116–1118.

Lee, N.H., Earle-Hughes, J., and Fraser, C. (1994) Agonist-mediated destabilization of m1 muscarinic acetylcholine receptor mRNA. Elements involved in mRNA stability are involved in the 3′-untranslated region. *J. Biol. Chem.* 269, 4291–4298.

Lefkowitz, R.J. (1993) G protein-coupled receptor kinases. *Cell* 74, 409–412.

Letourneur, F., and Klausner, R.D. (1992) A novel di-leucine motif and a tyrosine-based motif independently mediate lysosomal targeting and endocytosis of CD3 chains. *Cell* 69, 1143–1157.

Liggett, S.B., Freedman, N.J., Schwinn, D.A., and Lefkowitz, R.J. (1993) Structural basis for receptor subtype-specific regulation revealed by a chimeric β_3/β_2-adrenergic reeptor. *Proc. Natl. Acad. Sci. USA* 90, 3665–3669.

Lin, F.T., Krueger, K.M., Kendall, H.E., Daaka, Y., Fredericks, Z.L., Pitcher, J.A., and Lefkowitz, R.J. (1997) Clathrin-mediated endocytosis of the β-adrenergic receptor is regulated by phosphorylation/dephosphorylation of β-arrestin1. *J. Biol. Chem.* 272, 31051–31057.

Lohse, M.J. (1993) Molecular mechanisms of membrane receptor desensitization. *Biochim. Biophys. Acta* 1179, 171–188.

Lohse, M.J., Andexinger, S., Pitcher, J., Trukawinski, S., Codina, J., Faure, J.P., Caron, M.G., and Lefkowitz, R.J. (1992) Receptor-specific desensitization with purified proteins. Kinase dependence and receptor specificity of β-arrestin and arrestin in the β_2-adrenergic receptor and rhodopsin systems. *J. Biol. Chem.* 267, 8558–6564.

Lohse, M.J., Benovic, J.L., Caron, M.G., and Lefkowitz, R.J. (1990a) Multiple pathways of rapid β_2-adrenergic receptor desensitization. Delineation with specific inhibitors. *J. Biol. Chem.* 265, 3202–3209.

Lohse, M.J., Benovic, J.L., Codina, J., Caron, M.G., and Lefkowitz, R.J. (1990b) β-arrestin: A protein that regulates β-adrenergic receptor function. *Science* 248, 1547–1550.

Lohse, M.J., Krasel, C., Winstel, R., and Mayor, F. (1996) G-protein-coupled receptor kinases. *Kidney Int.* 49, 1047–1052.

Machida, C.A., Brunzow, J.R., Searles, R.P., Van Thol, H., Tester, B., Neve, K.A., Teal, P., Nipper, V., and Civelli, O. (1990) Molecular cloning and expression of the rat β_1 receptor gene. *J. Biol. Chem.* 265, 12960–12965.

Mahan, L.C., Koachman, A.M., and Insel, P.A. (1985) Genetic analysis of beta-adrenergic receptor internalization and down-regulation. *Proc. Natl. Acad. Sci. USA* 82, 129–133.

Mahan, L.C., McKernan, R.M., and Insel, P.A. (1987) Metabolism of alpha- and beta-adrenergic receptors in vitro and in vivo. *Annu. Rev. Pharmacol. Toxicol.* 27, 215–235.

Mak, J.C.W., Nishikawa, M., Shirasaki, H., Miyayasu, K., and Barnes, P.J. (1995) Protective effects of a glucocorticoid on downregulation of pulmonary β_2-adrenergic receptors in vivo. *J. Clin. Invest.* 96, 99–106.

Malbon, C.C., and Hadcock, R.J. (1988) Evidence that glucocorticoid response elements in the 5′-noncoding region of the hamster β_2-adrenergic receptor gene are obligate for glucocorticoid regulation of receptor mRNA levels. *Biochem. Biophys. Res. Commun.* 154, 676–681.

Malter, J.S. (1989) Identification of an AUUUA-specific messenger RNA binding protein. *Science* 246, 664–666.

McGraw, D.W., Jacobi, S.E., Hiller, F.C., and Cornett, L.E. (1996) Structural and functional analysis of the 5′-flanking region of the rat β_2-adrenergic receptor gene. *Biochim. Biophys. Acta* 1305, 135–138.

Münch, G., Dees, C., Hekman, M., and Palm, D. (1991) Multisite contacts involved in coupling of the β-adrenergic receptor with the stimulatory guanine-nucleotide-binding regulatory protein. *Eur. J. Biochem.* 198, 357–364.

Myer, V.E., Fan, X.C., and Steitz, J.A. (1997) Identification of HuR as a protein implicated in AUUUA-mediated mRNA decay. *EMBO J.* 16, 2130–2139.

Nahmias, C., Blin, N., Elalouf, J.M., Mattei, M.G., Strosberg, A.D., and Emorine, L.J. (1991) Molecular characterization of the mouse β_3-adrenergic receptor: Relationship with the atypical receptor of adipocytes. *EMBO J.* 10, 3721–3727.

Nakada, M.T., Haskell, K.M., Ecker, D.J., Stadel, J.M., and Crooke, S.T. (1989) Genetic regulation of β_2-adrenergic receptors in 3T3-L1 fibroblasts. *Biochem. J.* 260, 53–59.

Nantel, F., Bonin, H., Emorine, L.J., Zilberfarb, V., Strosberg, A.D., Bouvier, M., and Marullo, S. (1993) The human β_3-adrenergic receptor is resistant to short-term agonist-promoted desensitization. *Mol. Pharmacol.* 43, 548–555.

Neer, E.J., and Clapham, D.E, (1988) Roles of G protein subunits in transmembrane signalling. *Nature* 333, 129–134.

Neve, K.A., and Molinoff, P.B (1986) Turnover of β_1- and β_2-adrenergic receptors after down-regulation or irreversible blockade. *Mol. Pharmacol.* 30, 104–111.

O'Dowd, B.F., Hnatowich, M., Regan, J.W., Leader, W.M., Caron, M.G., and Lefkowitz, R.J. (1988) Site-directed mutagenesis of the cytoplasmatic domains of the human β_2-adrenergic receptor: Localization of regions involved in G protein-receptor coupling. *J. Biol. Chem.* 263, 15985–15992.

Okamoto, T., Murayama, Y., Hayashi, Y., Inagaki, M., Ogata, E, and Nishimoto, I. (1991) Identification of a G_s activator region of the β_2-adrenergic receptor that is autoregulated via protein kinaseA-dependent phosphorylation. *Cell* 67, 723–730.

Palczewski, K. (1994) Structure and functions of arrestins. *Protein Sci.* 3, 1355–1361.

Palczewski, K. (1997) GTP-binding-protein-coupled receptor kinases. *Eur. J. Biochem.* 248, 261–269.

Parola, A.L., and Kobilka, B.K. (1994) The peptide product of a 5′ leader cistron in the β_2-adrenergic receptor mRNA inhibits receptor synthesis. *J. Biol. Chem.* 269, 4497–4505.

Pende, A., Tremmel, K.D., DeMaria, C.T., Blaxall, B.C., Minobe, W.A., Sherman, J.A., Bisognato, J.D., Bristow, M.R., Brewer, G., and Port, J.D. (1996) Regulation of the mRNA-binding protein AUF1 by activation of the β-adrenergic receptor signal transduction pathway. *J. Biol. Chem.* 271, 8493–8501.

Peng, S.S.Y., Chen, C.Y.A., and Shyu, A.B. (1996) Functional characterization of a non-AUUUA AU-rich element from the c-*jun* proto-oncogene mRNA: Evidence for a novel class of AU-rich elements. *Mol. Cell. Biol.* 16, 1490–1499.

Perez, D.M., Piascik, M.T., Malik, N., Gaivin, R., and Graham, R.M. (1994) Cloning, expression, and tissue distribution of the rat homolog of the bovine α_{1C}-adrenergic receptor provide evidence for ist classification as the α_{1A} subtype. *Mol. Pharmacol.* 46, 823–831.

Pippig, S., Andexinger, S., Kiefer, D., Puzicha, M., Caron, M.G., Lefkowitz, R.J., and Lohse, M.J. (1993) Overexpression of β-arrestin und β-adrenergic receptor kinase augment desensitization of β_2-adrenergic receptors. *J. Biol. Chem.* 268, 3201–3208.

Pippig, S., Andexinger, S., and Lohse, M.J. (1995) Sequestration and recycling of β_2-adrenergic receptors permit receptor resensitization. *Mol. Pharmacol.* 47, 666–676.

Pitcher, J.A., Fredericks, Z.L., Stone, W.C., Premont, R.T., Stoffel, R.H., Koch, W.J., and Lefkowitz, R.J. (1996) Phosphatidylinositol 4,5-bisphosphate (PIP2)-enhanced G protein-coupled receptor kinase (GRK) activity: Location, structure and regulation of the PIP2 binding site distinguishes the GRK subfamilies. *J. Biol. Chem.* 271, 24907–24913.

Pitcher, J.A., Inglese, J., Higgins, J.B., Arriza, J.L., Casey, P.J., Kim, C., Benovic, J.L., Kwatra, M.M., Caron, M.G., and Lefkowitz, R.J. (1992a) Role of βγ subunits of G proteins in targeting the β-adrenergic receptor kinase to membrane bound receptors. *Science* 257, 1264–1267.

Pitcher, J.A., Lohse, M.J., Codina, J., Caron, M.G., and Lefkowitz, R.J. (1992b) Desensitization of the isolated β_2-adrenergic receptor by β-adrenergic receptor kinase, cAMP-dependent protein kinase, and protein kinase C occurs via distinct molecular mechanisms. *Biochemistry* 31, 3193–3197.

Pitcher, J.A., Payne, E.S., Csortos, C., DePaoli-Roach, A.A., and Lefkowitz, R.J. (1995) The G-protein-coupled receptor phosphatase: A protein phosphatase type 2A with a distinct subcellular distribution and substrate specificity. *Proc. Natl. Acad. Sci. USA* 92, 8343–8347.

Port, J.D., Huang, L.Y., and Malbon, C.C. (1992) β-adrenergic agonists that down-regulate receptor mRNA up-regulate a M_r 35,000 protein(s) that selectively binds to β-adrenergic receptor mRNAs. *J. Biol. Chem.* 267, 24103–24108.

Rajagopalan, L.E., and Malter, J.S. (1994) Modulation of granulocyte-macrophage colony-stimulating factor mRNA-stability *in vitro* by the adenosin-uridine binding factor. *J. Biol. Chem.* 269, 23882–23888.

Ramarao, C.S., Denker, J.M., Perez, D.M., Gaivin, R.J., Riek, R.P., and Graham, R.M. (1992) Genomic organization and expression of the human α_{1B}-adrenergic receptor. *J. Biol. Chem.* 267, 21936–21945.

Roesler, W.J., Vandenbark, G.R., and Hanson, R.W. (1988) Cyclic AMP and the induction of eukaryotic gene transcription. *J. Biol. Chem.* 263, 9063–9066.

Rohlff, C., Ahmad, S., Borellini, F., Lei, J., and Glazer, R.I. (1997) Modulation of transcription factor Spl by cAMP-dependent protein kinase. *J. Biol. Chem.* 272, 21137–21141.

Ross, J. (1995) mRNA stability in mammalian cells. *Microbiol. Rev.* 59, 423–450.

Roth, N.S., Campbell, P.T., Caron, M.G., Lefkowitz, R.J., and Lohse, M.J. (1991) Comparative rates of desensitization of β-adrenergic receptors by the β-adrenergic receptor kinase and the cyclic AMP-dependent protein kinase. *Proc. Natl. Acad. Sci. USA* 88, 6201–6204.

Rousseau, G., Nantel, F., and Bouvier, M. (1996) Distinct receptor domains determine subtype-specific coupling and desensitization phenotypes for human β_1- and β_2-adrenergic receptors. *Mol. Pharmacol.* 49, 752–760.

Rydelek-Fitzgerald, L., Li, Z., Machida, C.A., Fishman, P.H., and Duman, R.S. (1996) Adrenergic regulation of ICER (inducible cAMP early repressor) and β_1-adrenergic receptor gene expression in C6 glioma cells. *J. Neurochem.* 67, 490–497.

Sachs, A.B. (1993) Messenger RNA degradation in eukaryotes. *Cell* 74, 413–421.

Schiavi, S.C., Wellington, C.L., Shyu, A.B., Chen, C.Y.A., Greenberg, M.E., and Belasco, J.G. (1994) Multiple elements in the c-*fos* protein-coding region facilitate mRNA deadenylation and decay by a mechanism coupled to translation. *J. Biol. Chem.* 269, 3441–3448.

Schleicher, S., Boekhoff, I., Arriza, J., Lefkowitz, R.J., and Breer, H. (1993) A β-adrenergic receptor kinase-like enzyme is involved in olfactory signal termination. *Proc. Natl. Acad. Sci. USA* 90, 1420–1424.

Searles, R.P., Midson, C.N., Nipper, V.J., and Machida, C.A. (1995) Transcription of the rat β_1-adrenergic receptor gene. Characterization of the transcript and identification of important sequences. *J. Biol. Cem.* 270, 157–162.

Searles, R.P., Nipper, V.J., and Machida, C.A. (1994) The rhesus macaque β_1-adrenergic receptor gene: Structure of the gene and comparison of the flanking sequences with the rat β_1-adrenergic receptor gene. *DNA Sequence* 4, 231–241.

Seibold, A., January, B.G., Friedman, J., Hipkin, W., and Clark, R.B. (1998) Desensitization of β_2-adrenergic receptors with mutations of the proposed G protein-coupled receptor kinase phosphorylation sites. *J. Biol. Chem.* 273, 7637–7642.

Shaw, G., and Kamen, R. (1986) A conserved AU sequence from the 3′ untranslated region of GM-CSF mRNA mediates selective mRNA degradation. *Cell* 46, 659–667.

Shear, M., Insel, P.A., Melmon, K.M., and Coffino, P. (1976) Agonist-specific refractoriness induced by isoproterenol. Studies with mutant cells. *J. Biol. Chem.* 251, 7572–7576.

Shyu, A.B., Belasco, J.G., and Greenberg, M.E. (1991) Two distinct destabilizing elements in the c-*fos* message trigger deadenylation as a first step in rapid mRNA decay. *Genes. Dev.* 5, 221–231.

Sibley, D.R., Strasser, R.H., Benovis, J.L., Daniel, K., and Lefkowitz, R.J. (1986) Phosphorylation/dephosphorylation of the β-adrenergic receptor regulates its functional coupling to adenylyl cyclase and subcellular distribution. *Proc. Natl. Acad. Sci. USA* 83, 9408–9412.

Smale, S.T., and Baltimore, D. (1989) The "initiator" as a transcription control element. *Cell* 57, 103–113.

Snavely, M.D., Ziegler, M.G., and Insel, P.A. (1985) A new approach to determine rates of receptor appearance and disappearance *in vivo*. Application to agonist-mediated down-regulation of rat renal cortical β_1- and β_2-adrenergic receptors. *Mol. Pharmacol.* 27, 19–26.

Sterne-Marr, R., and Benovic, J.L. (1995) Regulation of G protein-coupled receptors by receptor kinases and arrestins. *Vitam. Horm.* 51, 193–234.

Stoffel, R.H., Pitcher, J.A., and Lefkowitz, R.J. (1997) Targeting G protein-coupled receptor kinases to their receptor substrates. *J. Membrane Biol.* 157, 1–8.

Strader, C.D., Sigal, I.S., Blake, A.D., Cheung, A.H., Register, R.B., Rands, E., Zemcik, B.A., Candelore, M.R., and Dixon, R.A.F. (1987) The carboxyl terminus of the hamster β-adrenergic receptor expressed in mouse L-cells is not required for receptor sequestration. *Cell* 49, 855–863.

Su, Y.F., Harden, T.K., and Perkins, J.P. (1980) Catecholamine-specific desensitization of adenylate cyclase. *J. Biol. Chem.* 255, 7410–7419.

Suzuki, T., Nguyen, C.T., Nantel, F., Bonin, H., Valiquette, M., Frielle, T., and Bouvier, M. (1992) Distinct regulation of β_1- and β_2-adrenergic receptors in chinese hamster fibroblasts. *Mol. Pharmacol.* 41, 542–548.

Thekkumkara, T.J., Thomas, W.G., Motel, T.J., and Baker, K.M. (1998) Functional role for the angiotensin II receptor (AT_{1A}) 3′-untranslated region in determining cellular responses to agonist: Evidence for recognition by RNA binding proteins. *Biochem. J.* 329, 255–264.

Tholanikunnel, B.G., Granneman, J.G., and Malbon, C.C. (1995) The M_r 35,000 β-adrenergic receptor mRNA-binding protein binds transcripts of G-protein-linked receptors which undergo agonist-induced destabilization. *J. Biol. Chem.* 270, 12787–12793.

Tholanikunnel, B.G., and Malbon, C.C. (1997) A 20-nucleotide (A + U)-rich element of β_2-adrenergic receptor (β_2AR) mRNA mediates binding to β_2AR-binding protein and is obligate for agonist-induced destabilization of receptor mRNA. *J. Biol. Chem.* 272, 11471–11478.

Thomas, R.F., Holt, B.D., Schwinn, D.A., and Liggett, S.B. (1992) Long-term agonist exposure induces upregulation of β_3-adrenergic receptor expression via multiple cAMP response elements. *Proc. Natl. Acad. Sci. USA* 89, 4490–4494.

Touhara, K., Koch, W.J., Hawes, B.E., and Lefkowitz, R.J. (1995) Mutational analysis of the pleckstrin homology domain of the β-adrenergic receptor kinase. Differential effects on Gβγ and phosphatidylinositol 4,5-bisphosphate binding. *J. Biol. Chem.* 270, 17000–17005.

Ungerer, M., Böhm, M., Elce, J.S., Erdmann, E., and Lohse, M.J. (1993) Altered expression of β-adrenergic receptor kinase and β_1-adrenergic receptors in the failing human heart. *Circulation* 87, 454–463.

Vakalopoulou, E., Schaack, J, and Shenk, T. (1991) A 32-kilodalton protein binds to AU-rich domains in the 3′ untranslated regions of rapidly degraded mRNAs. *Mol. Cell. Biol.* 11, 3355–3364.

van Spronsen, A., Nahmias, C., Krief, S., Briend-Sutren, M.M., Strosberg, A.D., and Emorine, L.J. (1993) The promoter and intron/exon structure of the human and mouse β_3-adrenergic-receptor genes. *Eur. J. Biochem.* 213, 1117–1124.

von Zastrow, M., and Kobilka, B.K. (1992) Ligand-regulated internalization and recycling of human β_2-adrenergic receptors between the plasma membrane and endosomes containing transferrin receptors. *J. Biol. Chem.* 267, 3530–3538.

von Zastrow, M., and Kobilka, B.K. (1994) Antagonist-dependent and -independent steps in the mechanism of adrenergic receptor internalization. *J. Biol. Chem.* 269, 18448–18452.

von Zastrow, M., Link, R., Daunt, D., Barsh, G., and Kobilka, B.K. (1993) Subtype-specific differences in the intracellular sorting of G protein-coupled receptors. *J. Biol. Chem.* 268, 763–766.

Wagner, B.J., DeMaria, C.T., Sun, Y., Wilson, G.M., and Brewer, G. (1998) Structure and genomic organization of the human AUF1 gene: Alternative pre-mRNA splicing generates four protein isoforms. *Genomics* 48, 195–202.

Wang, J., and Ross, E.M. (1995) The carboxyl-terminal anchorage domain of the turkey β_1-adrenergic receptor is encoded by an alternative spliced exon. *J. Biol. Chem.* 270, 6488–6495.

Wang, X., Nickenig, G., and Murphy, T.J. (1997) The vascular smooth muscle type I angiotensin II receptor mRNA is destabilized by cyclic AMP-elevating agents. *Mol. Pharmacol.* 52, 781–787.

Wellington, C.L., Greenberg, M.E., and Belasco, J.G. (1993) The destabilizing elements in the coding region of c-*fos* mRNA are recognized as RNA. *Mol. Cell. Biol.* 13, 5034–5042.

Winstel, R., Freund, S., Krasel, C., Hoppe, E., and Lohse, M.J. (1996) Protein kinase cross-talk: Membrane targeting of the β-adrenergic receptor kinase by protein kinase C. *Proc. Natl. Acad. Sci. USA* 93, 2105–2109.

Xu, N., Chen, C.Y.A., and Shyu, A.B. (1997) Modulation of the fate of cytoplasmatic mRNA by AU-rich elements: Key sequence features controlling mRNA deadenylation and decay. *Mol. Cell. Biol.* 17, 4611–4621.

Yu, S.S., Lefkowitz, R.J., and Hausdorff, W.P. (1993) β-adrenergic receptor sequestration: A potential mechanism of receptor resensitization. *J. Biol. Chem.* 268, 268, 337–341.

Zawel, L., and Reinberg, D. (1993) Initiation of transcription by RNA polymerase II: A multi-step process. *Progr. Nucl. Acid Res. Mol. Biol.* 44, 67–108.

Zhang, J., Barak, L.S., Winkler, K.E., Caron, M.G., and Ferguson, S.S.G. (1997) A central role for β-arrestins and clathrin-coated vesicle-mediated endocytosis in β_2-adrenergic receptor resensitizaion. Differential regulation of receptor resensitization in two distinct cell types. *J. Biol. Chem.* 272, 27005–27014.

Zhang, J., Ferguson, S.S.G., Barak, L.S., Menard, L., and Caron, M.G. (1996) Dynamin and β-arrestin reveal distinct mechanisms for G protein-coupled receptor internalization. *J. Biol. Chem.* 271, 18302–18305.

Zhang, W., Wagner, B.J., Ehrenman, K., Schaefer, A.W., DeMaria, C.T., Crater, D., DeHaven, K., Long, L. and Brewer, G. (1993) Purification, characterization, and cDNA cloning of an AU-rich element RNA-binding protein, AUF1. *Mol. Cell. Biol.* 13, 7652–7665.

Zubiaga, A.M., Belasco, J.G., and Greenberg, M.E. (1995) The nonamer UUAUUUAUU is the key AU-rich sequence motif that mediates mRNA degradation. *Mol. Cell. Biol.* 15, 2219–2230.

Editor-in-charge: Professor G. Schultz

Printing and binding: Druckerei Triltsch, Würzburg